AI-Enabled 6G Networks and Applications

AI-Enabled 6G Networks and Applications

Edited by

Deepak Gupta
Maharaja Agrasen Institute of Technology, India

Dr. Mahmoud Ragab
Associate Professor
Department of Information Technology
Faculty of Computing and Information Technology
King Abdulaziz University, Jeddah, Saudi Arabia

Romany Fouad Mansour
Associate Professor
Department of Mathematics
Faculty of Science
New Valley University, Egypt.

Aditya Khamparia
Babasaheb Bhimrao Ambedkar University, Satellite Centre,
Amethi, India

Ashish Khanna
Maharaja Agrasen Institute of Technology, India

This edition first published 2023
© 2023 John Wiley & Sons Ltd

The right of Deepak Gupta, Mahmoud Ragab, Romany Fouad Mansour, Aditya Khamparia, and Ashish Khanna to be identified as the author(s) of the editorial material in this work has been asserted in accordance with law.

Registered Office(s)
John Wiley & Sons, Inc., 111 River Street, Hoboken, NJ 07030, USA
John Wiley & Sons Ltd, The Atrium, Southern Gate, Chichester, West Sussex, PO19 8SQ, UK

Editorial Office
The Atrium, Southern Gate, Chichester, West Sussex, PO19 8SQ, UK

For details of our global editorial offices, customer services, and more information about Wiley products visit us at www.wiley.com.

Wiley also publishes its books in a variety of electronic formats and by print-on-demand. Some content that appears in standard print versions of this book may not be available in other formats.

Library of Congress Cataloging-in-Publication Data
Names: Gupta, Deepak, Ph.D., editor. | Ragab, Mahmoud, editor. | Mansour,
     Romany Fouad, editor. | Khamparia, Aditya, 1988– editor. | Khanna,
     Ashish (Of Guru Gobind Singh Indraprastha University), editor.
Title: AI-enabled 6G networks and applications / edited by Deepak Gupta,
     Mahmoud Ragab, Romany Fouad Mansour, Aditya Khamparia, Ashish Khanna.
Description: Hoboken, NJ : Wiley, 2023.
Identifiers: LCCN 2022023670 (print) | LCCN 2022023671 (ebook) | ISBN
     9781119812647 (cloth) | ISBN 9781119812708 (adobe pdf) | ISBN
     9781119812715 (epub)
Subjects: LCSH: Mobile communication systems. | Artificial intelligence.
Classification: LCC TK5103.2 .A434 2023 (print) | LCC TK5103.2 (ebook) |
     DDC 621.384–dc23/eng/20220812
LC record available at https://lccn.loc.gov/2022023670
LC ebook record available at https://lccn.loc.gov/2022023671

Cover image: © Krunja/Shutterstock
Cover design by Wiley

Set in 10.5/13pt STIXTwoText by Straive, Pondicherry, India
Printed and bound by CPI Group (UK) Ltd, Croydon, CR0 4YY

C9781119812647_241122

Contents

List of Contributors

Elsaid M. Abdelrahim
Department of Mathematics
Computer Science Division
Faculty of Science
Tanta University
Tanta, Egypt

Emad A.-B. Abdel-Salam
Department of Mathematics
Faculty of Science
New Valley University
El-Kharga, Egypt

Maha M. Althobaiti
Department of Computer Science
College of Computing and
Information Technology
Taif University
Taif, Saudi Arabia

Natasha Madera
Mechatronics Engineering
Program, Universidad Simón
Bolívar, Barranquilla, Colombia

Victor Hugo C. de Albuquerque
Department of Teleinformatics
Engineering
Federal University of Ceará
Fortaleza, Brazil

Marcello Carvalho dos Reis
Graduate Program in
Telecommunication Engineering
Federal Institute of Education
Science and Technology of Ceará
Fortaleza, Brazil

and

Meteora, Fortaleza, Brazil

Adnen El Amraoui
Univ. Artois, UR 3926 Laboratoire
de Génie Informatique
et d'Automatique de l'Artois
(LGI2A), F-62400, Béthune,
France

José Escorcia-Gutierrez
Research Center - CIENS, Escuela
Naval de Suboficiales ARC
Barranquilla, Barranquilla,
Colombia

and

Biomedical Engineering Program,
Corporación Universitaria
Reformada, Barranquilla,
Colombia

M. Ilayaraja
School of Computing
Kalasalingam Academy of
Research and Education
Krishnankoil
Tamilnadu, India

Ayman M. Mahmoud
Department of Mathematics
Faculty of Science
New Valley University
El-Kharga, Egypt

Vinita Malik
Graduate Program in
Telecommunication Engineering,
Federal Institute of Education,
Science and Technology of Ceará,
Fortaleza, Brazil

and

Central Library Central
University of Haryana,
Mahendragarh, Haryana, India

Romany F. Mansour
Department of Mathematics
Faculty of Science
New Valley University
El-Kharga, Egypt

Kanagaraj Narayanasamy
Department of Computer
Applications
Karpagam Academy of Higher
Education (Deemed to be
University)
Coimbatore, Tamilnadu, India

R. Pandi Selvam
PG Department of Computer
Science
Vidhyaa Giri College of Arts
and Science, Puduvayal
Karaikudi, Tamilnadu, India

Pooja Singh
Graduate Program in
Telecommunication Engineering,
Federal Institute of Education,
Science and Technology of Ceará,
Fortaleza, Brazil

and

Department of Computer Science
and Engineering, GL Bajaj
Institute of Technology and
Management, Greater Noida,
Uttar Pradesh, India

Carlos Soto
Mechanical Engineering Program
Universidad Autónoma del Caribe
Barranquilla, Colombia

Melitsa Torres-Torres
Research Group IET-UAC
Universidad Autónoma del Caribe
Barranquilla, Colombia

Preface

With the rapid development of diversified applications (e.g. virtual and augmented reality, telematics, remote surgery, and holographic projection) as well as construction of smart terminals and infrastructures, current networks (e.g. 4G and upcoming 5G networks) may not be able to completely meet quickly rising traffic demands. Keeping this in consideration both industry and academia have been engaged in research related to 6G networks. Recently, artificial intelligence (AI) has been utilized as a new paradigm for the design and optimization of 6G networks with a high level of intelligence. Therefore, this book presents an AI-enabled intelligent architecture, hardware, computing techniques, and related diversified applications for 6G networks to realize knowledge discovery, smart resource management, automatic network adjustment, and intelligent service provisioning with evolutionary computing techniques.

This book includes review of AI techniques for 6G networks and will focus on deployment of AI techniques to efficiently and effectively optimize the network performance, including AI-empowered mobile edge computing, intelligent mobility and handover management, and smart spectrum management. This book also includes the design of a set of evolutionary AI hybrid algorithms with communication protocols, showing how to use them in practice to solve problems relating to vehicular networks, aerial networks, and communication networks. It is intended as a reference guide for advanced hybrid computational intelligence methods for 6G supportive networks and protocols for graduate students and researchers in network forensics and optimization, computer science, and engineering.

Objective of the Book

The key features of the book are to highlight the encountered problems in 4G and 5G networks and provide suitable solutions to counter them. It also throws light on how to employ AI techniques to efficiently and effectively optimize the network performance and enable intelligent mobility. It enables usage of strong learning ability to achieve network intelligentization, closed-loop optimization, and intelligent wireless communication for 6G networks. The book also examines privacy issues and challenges related to data-intensive technologies in IoT enabled 6G networks. It also highlights important future research directions and potential solutions for AI-enabled intelligent 6G networks, including computation efficiency, algorithms robustness, hardware development, and energy management.

Organization of the Book

This book is organized in eight chapters with the following brief description:

Chapter 1: Metaheuristic Moth Flame Optimization Based Energy Efficient Clustering Protocol for 6G Enabled Unmanned Aerial Vehicle Networks
In proposed work author introduces a metaheuristic moth flame optimization algorithm for energy efficient clustering (MMFO-EEC) technique for 6G enabled unmanned aerial vehicle (UAV) networks. The major intention of the MMFO-EEC technique is the proficient election of cluster heads (CHs) and cluster organization in 6G enabled UAV networks. The presented MMFO-EEC technique mainly employs the MFO algorithm to effectually pick out the appropriate UAVs as CHs in the network.

Chapter 2: A Novel Data Offloading with Deep Learning Enabled Cyberattack Detection Model for Edge Computing in 6G Networks

This chapter develops a novel data offloading with deep learning enabled cyberattack detection (DADL-CAD) model for edge computing in 6G networks. The proposed DADL-CAD technique primarily designs recurrent neural network (RNN) model for traffic flow forecasting in the edge computing enabled 6G networks. The performance validation of the DADL-CAD technique is examined under various aspects, and the comparative study reported the supremacy of the DADL-CAD technique over the recent approaches.

Chapter 3: Henry Gas Solubility Optimization with Deep Learning Enabled Traffic Flow Forecasting in 6G Enabled Vehicular Networks

In this chapter author develops a novel Henry gas solubility optimization with deep learning enabled traffic flow forecasting (HSGODL-TFF) technique for 6G enabled vehicular networks. The presented HSGODL-TFF technique primarily intends to forecast the level of traffic in the 6G enabled VANET. HSGO algorithm can be applied for optimally modifying the hyperparameters (such as learning rate, epoch count, and batch size) of the DBN model thereby improving the forecasting performance. The experimental validation of the HSGODL-TFF model is performed on test data, and the results are inspected under several aspects.

Chapter 4: Crow Search Algorithm Based Vector Quantization Approach for Image Compression in 6G Enabled Industrial Internet of Things Environment

The proposed chapter introduces a novel crow search algorithm based vector quantization approach for image compression in 6G enabled IIoT environment, called CSAVQ-ICIIoT model. The proposed CSAVQ-ICIIoT model intends to accomplish effectual image compression by optimizing codebook construction process in 6G enabled IIoT platform. The CSAVQ-ICIIoT technique includes Linde–Buzo–Gray (LBG) with vector quantization (VQ) technique for image compression.

Chapter 5: Design of Artificial Intelligence Enabled Dingo Optimizer for Energy Management in 6G Communication Networks

This chapter presents an artificial intelligence enabled dingo optimizer for energy management (AIDO-EM) in 6G networks. The presented AIDO-EM technique involves the major goal of minimizing the energy utilization and maximizing the lifetime of the 6G enabled IoT devices. For accomplishing this, a new dingo optimization algorithm (DOA) is applied for cluster enabled routing to achieve effective data distribution among the devices and choose effective gateway heads (GWH).

Chapter 6: Adaptive Whale Optimization with Deep Learning Enabled RefineDet Network for Vision Assistance on 6G Networks

In this chapter an adaptive whale optimization with deep learning enabled RefineDet network (AWO-DLRDN) for vision assistance on 6G networks is discussed. The major intention of the AWO-DLRDN technique is to determine the nearby objects and their approximate distance to the visually impaired people. The proposed AWO-DLRDN technique primarily undergoes data augmentation and image annotation process as a preprocessing step. The performance validation of the AWO-DLRDN technique is experimented with using benchmark dataset, and the comparison study reported the enhancements of the AWO-DLRDN technique over the other techniques.

Chapter 7: Efficient Deer Hunting Optimization Algorithm Based Spectrum Sensing Approach for 6G Communication Networks

This chapter proposes a novel efficient deer hunting optimization algorithm based spectrum sensing approach (EDHO-SSA) for 6G communication networks. The presented EDHO-SSA technique mainly intends to manage the availability of spectrums that exist in the 6G networks. The EDHO-SSA technique is based on the hunting nature of the deers. It also derives an objective function to define the performance of SS including distinct parameters such as energy and throughput. The experimental result analysis of the EDHO-SSA technique is carried out, and the results are assessed with respect to various measures.

Chapter 8: Elite Oppositional Hunger Games Search Optimization Based Cooperative Spectrum Sensing Scheme for 6G Cognitive Radio Networks

In proposed work, elite oppositional hunger games search optimization based cooperative spectrum sensing (EOHGSO-CSS) scheme for 6G cognitive radio networks (CRNs) is discussed. The EOHGSO-CSS technique mainly intends to allocate the spectrum effectively in the 6G CRNs. The SS process can be carried out using different parameters such as interference, sensing time, threshold value, energy, throughput, and power allocation. Besides, the EOHSGO algorithm has been derived by the integration of elite oppositional based learning (EOBL) with traditional HGSO algorithm to improve its efficiency.

About the Editors

Dr. Deepak Gupta received a BTech degree in 2006 from the Guru Gobind Singh Indraprastha University, Delhi, India. He received ME degree in 2010 from Delhi Technological University, India, and PhD degree in 2017 from Dr. APJ Abdul Kalam Technical University (AKTU), Lucknow, India. He has completed his postdoc from National Institute of Telecommunications (Inatel), Brazil, in 2018. He has co-authored more than 183 journal articles including 145 SCI papers and 44 conference articles. He has authored/edited 50 books, published by IEEE-Wiley, Elsevier, Springer, Wiley, CRC Press, De Gruyter, and Katsons. He has filled four Indian patents. He is convener of ICICC, ICDAM, DoSCI, and ICCCN Springer conferences series. Currently he is associate editor of *Alexandria Journal* (Elsevier), *Expert Systems* (Wiley), and *Intelligent Decision Technologies* (IOS Press). He is the recipient of 2021 IEEE System Council Best Paper Award. He has been featured in the list of top 2% scientist/researcher database in the world [Table-S7-singleyr-2019]. He is also working toward promoting startups and also serving as a startup consultant. He is also a series editor of "Elsevier Biomedical Engineering" at Academic Press, Elsevier; "Intelligent Biomedical Data Analysis" at De Gruyter, Germany; and "Explainable AI (XAI) for Engineering Applications" at CRC Press. He is appointed as a consulting editor at Elsevier.

Dr. Mahmoud Ragab obtained his PhD degree in sorting algorithms and weighted branching processes from the faculty of Mathematics and Natural Sciences of the Christian-Albrechts-University at Kiel (CAU), Schleswig-Holstein, Germany. He is an associate professor at Department of Information Technology, Faculty of Computing and Information Technology, King Abdulaziz University, Jeddah, Saudi Arabia. He is currently the head of the Biological Quantum Computing Unit in the Centre for Artificial Intelligence in Precision Medicine, King Abdulaziz University. He is currently a consultant for the Vice Presidency for Graduate Studies and Scientific Research and a consultant for Vice President for Development, at KAU as well. Also, he is an associate professor for Data Science at the Mathematics Department, Faculty of Science, Al-Azhar University, Cairo, Egypt. In addition, Dr. Mahmoud is working in different research groups at many universities such as the Combinatorial Optimization and Graph Algorithms Group (COGA), Faculty II Mathematics and Natural Sciences, Berlin University of Technology, Berlin Germany; Faculty of Informatics and Computer Science, British University in Egypt (BUE), Cairo, Egypt; Arbeitsgruppe Stochastik, faculty of Mathematics and Natural Sciences, Christian-Albrechts-University at Kiel CAU, Kiel, Germany; and Department of Computer Science and Automation, Integrated Communication Systems Group, Ilmenau University of Technology TU Ilmenau, Thuerengen, Germany. His research focuses on artificial intelligence in biomedical applications, deep learning, sorting algorithms/efficiency, optimization, mathematical modeling, data science/data analysis, mathematical/applied statistics, neural networks, time series analysis, and quantum computation.

Dr. Romany Fouad Mansour received his PhD in Computer Science in 2009, from Assiut University, Egypt. He is currently working as an associate professor at Department of Mathematics, Faculty of Science, New Valley University, Egypt. His research interests include artificial intelligence, pattern recognition, computer vision, computer networks, soft computing, image processing, bioinformatics, and evolutionary computation.

Dr. Aditya Khamparia has expertise in teaching, entrepreneurship, and Research & Development of eight years. He is currently working as assistant professor and coordinator of Department of Computer Science, Babasaheb Bhimrao Ambedkar University, Satellite Centre, Amethi, India. He received his PhD degree from Lovely Professional University, Punjab, in May 2018. He has completed his MTech from VIT University and BTech from RGPV, Bhopal. He has completed his PDF from UNIFOR, Brazil. He has around 95 research papers along with book chapters including more than 15 papers in SCI indexed journals with cumulative impact factor of above 50 to his credit. Additionally, he has authored and edited five books. Furthermore, he has served the research field as a keynote speaker/session chair/reviewer/TPC member/guest editor and many more positions in various conferences and journals. His research interest includes machine learning, deep learning, educational technologies, and computer vision.

Dr. Ashish Khanna [M'19, SM'20] has expertise in teaching, entrepreneurship, and Research & Development with specialization in subjects of computer science engineering. He received his PhD degree from National Institute of Technology, Kurukshetra, in March 2017. He has completed his PDF from Internet of Things Lab at Inatel, Brazil. He completed his MTech in 2009 and BTech from GGSIPU, Delhi, in 2004. He is part of AD Scientific World ranking report as a leading researcher. He has around 160 accepted and published research papers and book chapters in reputed SCI, Scopus journals, conferences, and reputed book series including 82 papers accepted and published in SCI indexed journals and has a cumulative impact factor of above 300. He also has five published patents to his credit. Additionally, he has co-authored, edited, and is currently editing around 37 books. He is also serving as a series editor in publishing houses like De Gruyter (Germany) of "Intelligent Biomedical Data Analysis" series, Elsevier of "Intelligent Biomedical Data Analysis," and CRC Press of "Intelligent Techniques in Distributed Systems: Principles and Applications." He is also acting as a consulting editor for Elsevier. His research interests include distributed systems and its variants (MANET, FANET, VANET,

IoT), machine learning, NLP, and many more. He is the recipient of the 2021 IEEE System Council Best Paper Award. He is serving as convener/ general chair in some Springer International Conferences series like ICICC, ICDAM, DoSCI, ICCCN, and many more. He is a senior IEEE member (SMIEEE) and an ACM member too. He has played a key role in promoting and initiating several startups and is also a startup consultant. He has also played a key role in promoting Smart India Hackathon at MAIT. He initiated the first of its kind event in India "WHERE STARTUP MEETS INVESTOR" in collaboration with Universal Inovator and SIIF SSCBS, DU, under the banner of ICICC-2020 international conference. First time a research based conference was conducted along with a startup funding event. He also played a key role in India's first conference on Patents (ICIIP-2021) in association with UI and IGDTUW Delhi. He is serving as a consultant and mentor to some of the successful startups. He is also a key originator of Bhavya Publication house and Universal Inovator, an Indian research lab. His Google scholar index is 4000.

1

Metaheuristic Moth Flame Optimization Based Energy Efficient Clustering Protocol for 6G Enabled Unmanned Aerial Vehicle Networks

Adnen El Amraoui

Univ. Artois, UR 3926 Laboratoire de Génie Informatique et d'Automatique de l'Artois (LGI2A), F-62400, Béthune, France

1.1 Introduction

The sixth generation (6G) wireless network is the emerging networking technology across the globe that will give pervasive availability. This arising innovation is described by a few highlights contrasted with the past networking advances such as holographic profound correspondence, artificial intelligence (AI), visible light communication, 3D inclusion edge, and ground and ethereal wireless focal points for cloud usefulness [1]. One of the most significant highlights is the finished dependence on AI with its different advancements to control such an enormous networking engineering. Figure 1.1 shows the application area of 6G network. Wireless correspondence frameworks have encountered significant progressive advancement throughout the most recent years [2]. With the quick advancement of 3GPP 5G stage 2 normalization, the business organization of 5G applications being sent all around the world cannot completely address the difficulties brought by the fast increment of traffic and the constant prerequisite of administrations [3, 4].

AI-Enabled 6G Networks and Applications, First Edition. Edited by Deepak Gupta, Mahmoud Ragab, Romany Fouad Mansour, Aditya Khamparia, and Ashish Khanna.
© 2023 John Wiley & Sons Ltd. Published 2023 by John Wiley & Sons Ltd.

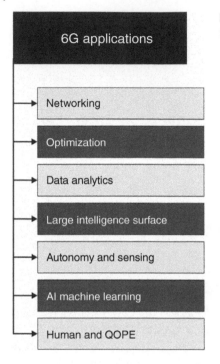

Figure 1.1 Applications of 6G network.

For this benefit, industry and the scholarly world are now learning and exploring the sixth era (6G) correspondence frameworks. Computer based intelligence, as a feature of AI, includes encouraging the machines to perform errands freely founded on settling on information-driven choices. Human-made intelligence can precisely assess different boundaries and backing intelligent independent direction [5]. The difficulties faced in applying AI strategies for 6G wireless networks is being examined. The use of AI methods in 6G wireless correspondence frameworks has been the subject that has gained interest as of late.

Unmanned aerial vehicle (UAV) node comprises of a radio handset (play the roles of transmitter and beneficiary), a microcontroller, and an electronic circuit for interacting with the related sensors and an energy source (normally a battery or an installed type of energy gathering) [6]. In several applications, sensors node requests self-association as a consequence of the arbitrariness existing in non-deterministic, uncontrolled geographies. Contingent on the idea of uses, different classes of sensor nodes are accommodated observing boundaries such as dampness, temperature, movement of articles, and sound. [7]. Generally, sensor networks make up for human endeavors in blocked off territories and present more agreeable, brilliant depictions of the climate. In the new future, sensor networks would overcome a basic piece of human existence and make current PC, versatile specified gadgets, and rest of the registering gadget less famous [8]. The lifetime of a sensor network is staunchly dependent upon the energy use, especially when there is no preparation for human admittance to the elaborate sensory nodes. Thus, numerous techniques have been proposed to limit energy utilization in UAVs. The design of UAV network displays many difficulties according to this point of view [9].

Topology management is viewed as a suitable method to guarantee steady, dependable, reliable, and productive network foundations in impromptu networks like UAVs. Clustering is one of the most well-known methods for UAV topology handling. A clustering method arranges nodes into a group of gatherings termed cluster in light of a group of predetermined measures, for example, supports quality of service (QoS), advance asset use, and network load adjusting. [10]. The cluster has minimum one cluster head (CH) that collects information from distinct nodes in the cluster called individual and straightforwardly transmit the (combined) information to binary signals (BS), or by implication utilizing different nodes called brokers nodes. With these clustering methods, asset compelled node does not need to send information to passages that (sink) can cause energy exhaustion, asset utilization failure, and obstruction.

Arafat and Moh [11] presented a bioinspired localization (BIL) and clustering (BIC) scheme for UAV network in order to detect and monitor wildfire. Primarily, a hybrid gray wolf optimizer (HGWO) with energy efficient two-dimensional BIL scheme is developed depending upon the HGWO, to reduce localization error and increase accuracy. Ma et al. [12] proposed a coordinate based optimizer, integrating genetic algorithm (GA) and clustering approach for solving multiple task allocation and path planning processes. It effectually computes the number of UAVs fulfilling the constraints and determining the optimal flight route in the UAV network. In [8], a clustering model has been designed for speeding up the speech at which the multi-UAV formation converged. With the consideration of the flight control factor for improving the convergence of multi-UAV, the UAV creates a flock. For accomplished secured data transmission, hierarchical virtual communication ring (HVCR) model has been developed.

This study introduces a metaheuristic moth flame optimization algorithm for energy efficient clustering (MMFO-EEC) technique for 6G enabled UAV networks. The presented MMFO-EEC technique mainly employs the MFO algorithm to effectually pick out the appropriate UAVs as CHs in the network. Besides, the MMFO-EEC technique derives a fitness function comprising distinct input parameters for accomplishing improved network performance. A wide range of simulations were carried out to highlight the enhancements of the MMFO-EEC technique, and the experimental values are inspected under several measures.

1.2 The Proposed Model

In this study, a new MMFO-EEC technique has been developed for the proficient election of CHs and cluster organization in 6G enabled UAV networks. The presented MMFO-EEC technique applied the MFO algorithm to effectually pick out the appropriate UAVs as CHs in the network. Besides, the MMFO-EEC technique derives a fitness function comprising distinct input parameters for accomplishing improved network performance.

1.2.1 Network Model

Assume a UAV network containing various UAVs. The UAV employs smaller to medium-sized drone. Moreover, a simple collision method is utilized for avoiding collision whereby the altitude of UAV gets modified. Also, the UAV moves at a speed of 30 m/s. Each UAV depends on position aware element, and the features of position aware method enable accurate and effective clustering procedure. Generally, the position data that can be accomplished from external system includes global positioning service (GPS). Here, it is noted that the inertial measurement unit and GPS is provided to location and motion sensing of the UAV. Each UAV knows their position, position of the ground station, and position of the neighboring UAV. In addition, the UAV is armed with shorter and longer rage transmission with 6G techniques. The previous one is utilized for transmitting information within the cluster, named intra-cluster transmission where the last one is applied for transmitting information among the ground station and CHs. Besides, in UAV-related mobile transmission, it aims at an explicit data rate R_b with a predetermined modulation approach and quadrature phase shift keying (QPSK). Figure 1.2 provides an overview of UAV network.

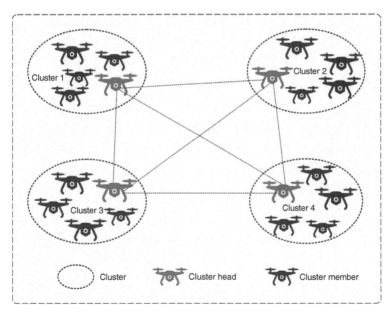

Figure 1.2 Overview of UAV network.

1.2.2 Algorithmic Procedure of MFO Algorithm

The MFO approach is a population based metaheuristic model that simulates moth behavior in the night around the flame. Flames and moths are the major components of the MFO approach. During the night time, the moth flies around the flame at a certain angle. When the moth sees the light source, it continues to fly in a straight line toward the light source. When the moth approaches the light source, it moves around the light source in a spiral path [13]. Moth is the searching agent, and the flame is the optimal location established until now. Thus, each population takes location as a solution. It can be expressed as follows:

$$
M = \begin{bmatrix} M_{1,1} & M_{1,d} \\ \vdots & \vdots \\ M_{n,1} & M_{n.d} \end{bmatrix} \tag{1.1}
$$

Where n represents the amount of moths and d indicates the amount of problem parameters. The matrix OM is other matrix to save the fitness value of the solution that illustrates the level of solution quality [14].

$$
OM = [OM_1 \; OM_2 \; ... \; OM_n]^T \tag{1.2}
$$

Furthermore, the flame is another component of the MFO approach. Matrix F representing the flame is demonstrated in the following:

$$
F = \left\{ \begin{array}{cc} F_{1,1} & F_{1,d} \\ \vdots & \vdots \\ F_{n,1} & F_{n.d} \end{array} \right\} \tag{1.3}
$$

Here, n characterizes the amount of moths (or flames), and d indicates the dimension or the amount of parameters of the problem. Note that the dimension of M and F matrices are equivalent to one another. The matrix OF has the fitness value for the flame:

$$
OF = [OF_1 \; OF_2 \; ... \; OF_n]^T \tag{1.4}
$$

In fact, flames and moths show solution. All the moths search the space around its flame and all the iterations discover an optimal solution, and the flame illustrates the optimal solution. The MFO process employs three functions for initializing the arbitrary locations of the moth (I), moves the moth in the searching space (P), and terminates the searching condition (T) as follows:

$$MFO = (I, P, T) \tag{1.5}$$

Where I represents a function that initializes the population initialization of the moth.

$$I : \emptyset \rightarrow \{M, OM\} \tag{1.6}$$

Also, P represents a function that moves the moth as per Eq. (1.6):

$$P : M \rightarrow M \tag{1.7}$$

The final function utilized is the function T. When the end criteria is fulfilled, T returns True, and when the end condition is not fulfilled, T returns False as follows:

$$T : M - \rightarrow \{\text{true, false}\} \tag{1.8}$$

Flames and moths are the building blocks of the MFO approach. Moth flies around the searching space, whereas the flame shows the optimal location. Moth flies around the flame and upgrades the position by discovering optimal location.

1.2.3 Design of MMFO-EEC Technique

The aim of MMFO-EEC technique is to separate the *n UAV nodes* into a predetermined or optimal amount of clusters C_{opt}. During clustering process, the neighboring node is allocated to the CH through Euclidean distance that makes minimal transmission range leading to a decrease in energy utilization. However, it is difficult to find the distance in extremely mobile situation. To resolve the problem, the distance to adjoining UAV is defined using MMFO-EEC technique. To choose CH and create cluster, the MMFO-EEC technique considers the problem as a maximization problem and derives fitness functions including average distance to neighboring UAVs (*DTN*), UAV degree (*DEG*), and residual energy level (*REL*).

$$F(i) = \alpha \times REL + \beta \times ADTN + \gamma \times DEG, \tag{1.9}$$

Where $\alpha + \beta + \gamma = 1$. The expansion of UAV($x$) in k-bit data to receive UAY(y) is positioned at distance d is as follows:

$$REL = E - \left(E_T(k, d) + E_{R(k)}\right) \tag{1.10}$$

In which E signifies the existing energy level of UAV and E_T shows the energy consumed for transmitting information.

$$E_T(k, d) = kE_e + KE_a d^2 \tag{1.11}$$

Here E_e indicates the energy of electron and E_a denotes the important amplified energy, and $E_{R(k)}$ shows the energy expended for data reception as follows:

$$E_{R(k)} = kE_e \tag{1.12}$$

As well, the advanced depay tolerant network (ADTN) signifies the normal value of distance of the adjoining UAVs in its 1-hop broadcast range [15].

$$ADTN = \frac{\sum_{j=1}^{NB_i} dist\left(i, nb_j\right)}{NB_i}, \tag{1.13}$$

While $dist(i, nb_j)$ shows the distance from the UAV to the nearer jth UAV.

In time t, the DEG characterizes the UAV degree indicating the amount of adjacent node existing for the UAV,

$$DEG = |N(x)| \tag{1.14}$$

Here $N(x) = \{n_y/dist(x, y) < trans_{range}\}x \neq y$, and $dist(x, y)$ means the distance among two UAVs n_x and n_y, $trans_{range}$ denotes the broadcast range of the UAV.

1.3 Experimental Validation

The experimental result analysis of the MMFO-EEC model has been validated under varying UAV count. Figure 1.3 offers a comparison study of the MMFO-EEC model with recent methods under distinct UAVs. The results indicated that the MMFO-EEC model has accomplished least energy capture model (ECM) over the existing approaches. For instance, with 10 UAVs, the MMFO-EEC model has obtained minimum ECM of 71.21 mJ, whereas the MEEDG-CUAV, SOCS, BICSF, and EALC models have attained maximum ECM of 87.32, 96.47, 96.47, and 100.87 mJ, respectively. Moreover, with 20 UAVs, the MMFO-EEC model has reached lower ECM of 92.32 mJ, whereas the MEEDG-CUAV, SOCS, BICSF, and EALC models have achieved higher ECM of 108.68, 120.63, 125.81, and 127.82 mJ, respectively. Furthermore, with 100 UAVs, the MMFO-EEC model has gained least ECM of 194.13 mJ, whereas the MEEDG-CUAV, SOCS, BICSF, and EALC models have accomplished increased ECM of 210, 230.63, 252.57, and 269.40 mJ, respectively.

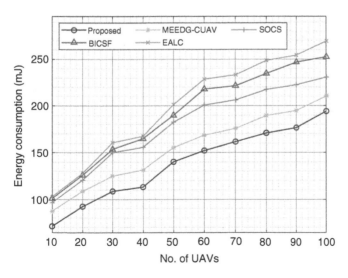

Figure 1.3 Comparative ECM analysis of MMFO-EEC with recent models.

Figure 1.4 provides a comprehensive NWLT study of the MMFO-EEC model with existing techniques under diverse UAVs. The experimental values reported that the MMFO-EEC model has resulted in enhanced NWLT over the existing approaches. For instance, with 10 UAVs, the MMFO-EEC model has depicted increased NWLT of 5990 rounds, whereas the MEEDG-CUAV, SOCS, BICSF, and EALC models have obtained decreased NWLT of 5920, 5570, 5520, and 5330 rounds, respectively. Besides, with 20 UAVs, the MMFO-EEC model has reached increased NWLT of 5960 rounds, whereas the MEEDG-CUAV, SOCS, BICSF, and EALC models have achieved decreased NWLT of 5810, 5520, 5290, and 5150 rounds, respectively. Additionally, with 100 UAVs, the MMFO-EEC model has gained improved NWLT of 4520 rounds, whereas the MEEDG-CUAV, SOCS, BICSF, and EALC models have resulted in enhanced NWLT of 4160, 3550, 3380, and 3280 rounds, respectively.

Table 1.1 and Figure 1.5 illustrate a detailed therapeutic (THRP) study of the MMFO-EEC model with existing techniques under diverse UAVs.

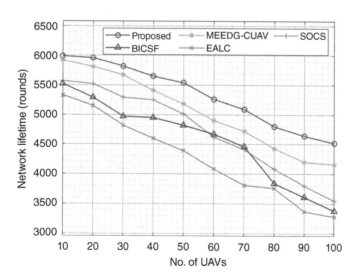

Figure 1.4 NWLT Analysis of MMFO-EEC with recent models.

Table 1.1   THRP analysis of MMFO-EEC with recent models.

No. of UAVs	Throughput (Mbps)				
	Proposed	**MEEDG-CUAV**	**SOCS**	**BICSF**	**EALC**
10	99.33	98.90	93.21	91.98	89.81
20	98.61	96.21	86.80	81.09	77.22
30	94.81	90.24	79.98	71.92	67.09
40	91.54	87.02	72.20	66.22	60.00
50	90.36	84.95	67.88	60.80	56.20
60	88.56	83.04	64.08	58.29	54.11
70	85.66	81.22	61.93	55.06	53.11
80	83.26	77.82	61.29	52.92	51.86
90	81.40	74.98	59.30	50.08	51.14
100	78.73	74.14	56.90	48.93	48.01

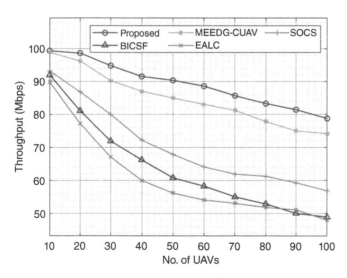

Figure 1.5   Comparative THRP analysis of MMFO-EEC with recent models.

The experimental values reported that the MMFO-EEC model has resulted in enhanced THRP over the existing approaches. For instance, with 10 UAVs, the MMFO-EEC model has depicted increased THRP of 99.33 Mbps, whereas the MEEDG-CUAV, SOCS, BICSF, and EALC models have obtained decreased THRP of 98.90, 93.21, 91.98, and 89.81 Mbps, respectively. Besides, with 20 UAVs, the MMFO-EEC model has reached increased THRP of 98.61 Mbps, whereas the MEEDG-CUAV, SOCS, BICSF, and EALC models have achieved decreased THRP of 96.21, 86.80, 81.09, and 77.22 Mbps, respectively. Additionally, with 100 UAVs, the MMFO-EEC model has gained improved THRP of 78.73 Mbps, whereas the MEEDG-CUAV, SOCS, BICSF, and EALC models have resulted in enhanced THRP of 74.14, 56.90, 48.93, and 48.01 Mbps, respectively

Table 1.2 and Figure 1.6 offer a comparison average delay (ADE) study of the MMFO-EEC model with recent methods under distinct UAVs. The results indicated that the MMFO-EEC model has accomplished

Table 1.2 Result analysis of existing with proposed method in terms of average delay (seconds).

No. of UAVs	Average delay (s)				
	Proposed	MEEDG-CUAV	SOCS	BICSF	EALC
10	3.70	4.26	4.45	4.38	4.46
20	3.99	4.45	4.96	5.06	6.12
30	4.30	4.82	5.60	5.87	7.20
40	5.62	6.10	6.46	7.24	7.85
50	6.37	6.92	8.00	8.42	8.65
60	7.61	8.02	8.63	8.90	9.24
70	7.80	8.29	9.06	9.64	9.84
80	8.08	8.57	9.35	9.94	10.70
90	8.67	9.08	9.89	10.91	11.24
100	8.65	9.14	10.51	11.14	11.40

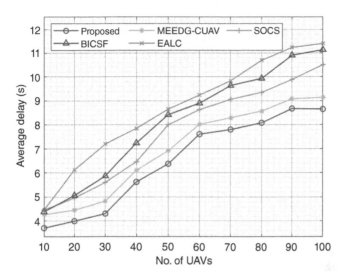

Figure 1.6 Comparative ADE analysis of MMFO-EEC with recent models.

least ADE over the existing approaches. For instance, with 10 UAVs, the MMFO-EEC model has obtained minimum ADE of 3.70 seconds, whereas the MEEDG-CUAV, SOCS, BICSF, and EALC models have attained maximum ADE of 4.26, 4.45, 4.38, and 4.46 seconds, respectively. Moreover, with 20 UAVs, the MMFO-EEC model has reached lower ADE of 3.99 seconds, whereas the MEEDG-CUAV, SOCS, BICSF, and EALC models have achieved higher ADE of 4.45, 4.96, 5.06, and 6.12 seconds, respectively. Furthermore, with 100 UAVs, the MMFO-EEC model has gained least ADE of 8.65 seconds, whereas the MEEDG-CUAV, SOCS, BICSF, and EALC models have accomplished increased ADE of 9.14, 10.51, 11.14, and 11.40 seconds, respectively.

1.4 Conclusion

In this study, a new MMFO-EEC technique has been developed for the proficient election of CHs and cluster organization in 6G enabled UAV networks. The presented MMFO-EEC technique applied the MFO algorithm to effectually pick out the appropriate UAVs as CHs in the network. Besides, the MMFO-EEC technique derives a fitness function comprising of the distinct input parameters for accomplishing improved network performance. A wide range of simulations were carried out to highlight the enhancements of the MMFO-EEC technique, and the experimental values reported the improved performance of the MMFO-EEC technique over the recent approaches. Therefore, the MMFO-EEC technique can be applied as a proficient tool for 6G enabled UAV networks. In future, data aggregation schemes can be included at the CHs for improved network performance.

References

1 Na, Z., Liu, Y., Shi, J. et al. (2021). UAV-supported clustered NOMA for 6G-enabled Internet of Things: trajectory planning and resource allocation. *IEEE Internet of Things Journal* 8 (20): 15041–15048.

2 Spyridis, Y., Lagkas, T., Sarigiannidis, P. et al. (2021). Towards 6G IoT: tracing mobile sensor nodes with deep learning clustering in UAV networks. *Sensors* 21 (11): 3936.

3 Yang, F., Song, J., Xiong, W., and Cui, X. (2021). UAV-based collaborative electronic reconnaissance network for 6G. *Wireless Communications and Mobile Computing* 2021: 1–7.

4 Strinati, E.C., Barbarossa, S., Choi, T. et al. (2020). 6G in the sky: in-demand intelligence at the edge of 3D networks. *arXiv* 10: 312–324.

5 Liu, R., Liu, A., Qu, Z., and Xiong, N.N. (2021). An UAV-enabled intelligent connected transportation system with 6G communications for internet of vehicles. *IEEE Transactions on Intelligent Transportation Systems.* 23: 1–15.

6 Chang, H., Wang, C.X., Liu, Y. et al. (2020). A novel nonstationary 6G UAV-to-ground wireless channel model with 3-D arbitrary trajectory changes. *IEEE Internet of Things Journal* 8 (12): 9865–9877.

7 Kovalenko, V., Alzaghir, A., Volkov, A. et al. (2020). Clustering algorithms for UAV placement in 5G and beyond networks. *2020 12th International Congress on Ultra Modern Telecommunications and*

Control Systems and Workshops (ICUMT), pp. 301–307, Brno, Czech Republic (14 October 2020). UK: IEEE.

8 Wu, J., Zou, L., Zhao, L. et al. (2019). A multi-UAV clustering strategy for reducing insecure communication range. *Computer Networks* 158: 132–142.

9 Arafat, M.Y. and Moh, S. (2019). Localization and clustering based on swarm intelligence in UAV networks for emergency communications. *IEEE Internet of Things Journal* 6 (5): 8958–8976.

10 Bupe, P., Haddad, R., and Rios-Gutierrez, F. (2015). Relief and emergency communication network based on an autonomous decentralized UAV clustering network. *SoutheastCon 2015*, pp. 1–8, UK (12–14 September 2015). UK: IEEE.

11 Arafat, M.Y. and Moh, S. (2021). Bio-inspired approaches for energy-efficient localization and clustering in UAV networks for monitoring wildfires in remote areas. *IEEE Access* 9: 18649–18669.

12 Ma, Y., Zhang, H., Zhang, Y. et al. (2019). Coordinated optimization algorithm combining GA with cluster for multi-UAVs to multi-tasks task assignment and path planning. *2019 IEEE 15th International Conference on Control and Automation (ICCA)*, pp. 1026–1031, Edinburgh, UK (14 November 2019). UK: IEEE.

13 Mirjalili, S. (2015). Moth-flame optimization algorithm: a novel nature-inspired heuristic paradigm. *Knowledge-Based Systems* 89: 228–249.

14 Pelusi, D., Mascella, R., Tallini, L. et al. (2020). An improved moth-flame optimization algorithm with hybrid search phase. *Knowledge-Based Systems* 191: 105277.

15 Almasoud, A.S., Ben, S., Nemri, N. et al. (2022). Metaheuristic based data gathering scheme for clustered UAVs in 6G communication network. *CMC-Computers, Materials & Continua* 71 (3): 5311–5325.

2

A Novel Data Offloading with Deep Learning Enabled Cyberattack Detection Model for Edge Computing in 6G Networks

Elsaid M. Abdelrahim

Department of Mathematics Computer Science Division, Faculty of Science, Tanta University, Tanta, Egypt

2.1 Introduction

Since 5G has brought vertical transformation for changing the society and progressively opened the curtain of internet of everything, 6G has opened an innovative era of "Internet of Intelligence" with connected things, intelligence, and people, resolving human problems in various aspects [1]. To authorize the 6G network using artificial intelligence (AI) abilities, massive amount of multi-modal information (for example, user videos, behavior records, and audios) of physical environments would be generated endlessly by the Internet of Things (IoT) and mobile gadgets that reside at the network edge. Driving by this trend, there is urgency for pushing the AI frontier to the network edge for completely unleashing the 6G network [2]. To satisfy the user need, edge computing, an emergent model that drives computational services and tasks from the network core to the network edge, was extensively known as essential component for the forthcoming 6G network. Mobile edge computing (MEC) is an alternative transmission system that provides service to users. This tool facilitates a mobile computing model with autonomous server and function that is attained from IoT or cloud

AI-Enabled 6G Networks and Applications, First Edition. Edited by Deepak Gupta, Mahmoud Ragab, Romany Fouad Mansour, Aditya Khamparia, and Ashish Khanna.

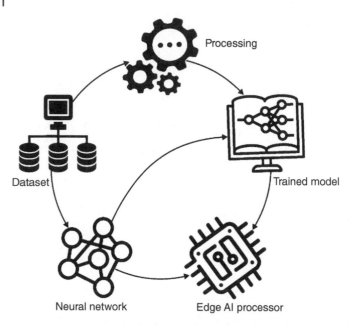

Figure 2.1 Components of AI in edge computing.

layers. The derivative resource is extended for users to provide complex services [3, 4]. The facility includes mobility, support transformation, elasticity, reliability, and adaptability characteristics that support varying user device density. Figure 2.1 illustrates the modules of AI from edge computing.

The computing request is streamlining from the edge layer to top layer including IoT or cloud [5]. Different applications and facilities associated with fog and edge frameworks are taken into account as part of IoT to generate scalable transmission networks with data analytics and computations [6]. The task of data offloading from the end users to MEC for more computation is studied broadly from existing studies when exploring the transmission and computational limits [7]. Numerous approaches and methods to computation offloading in MEC were technologically advanced from the research to assign energy utilization, decrease the computation latency, and radio resource effectively [8].

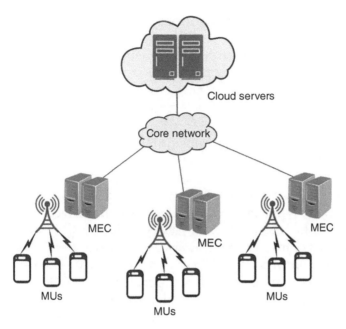

Figure 2.2 Structure of MEC.

Achieving better results in dynamic and complex multiple users wireless MEC schemes are complicated procedures. In addition, the security threat met in forwarding information are tackled in every offloading method [9]. Besides, inadequate data security checks may rapidly surpass the benefit of MEC technique. In order to counter the cyber threat in MEC, it can be critical to distinguish earlier cyberattacks, thus performing faster countermeasure to avoid the risk [10]. Figure 2.2 shows the infrastructure of MEC.

The researchers in [11] presented a mathematical method of the overall service delay of cybertwin based MEC scheme, which comprises migration of virtual servers, numerous physical servers, user mobility at distinct network tiers, content request/caching, processing, backhaul, and fronthaul communication as well as presented an algorithm to guide the process of cybertwins and the control plane from MEC situation. Liao et al. [12] presented a distributed two stage offloading (DTSO) method for providing trade-off solution. Initially, by presenting the queuing concept and considering channel interference, a combinatory optimization issue is generated for calculating the offloading possibility of all the stations. Next, the novel issue is transformed into nonlinear optimized issue that is resolved with sequential quadratic programming (SQP) model.

In Ref. [13], MEC scheme using a constrained computational buffer at the edge server is taken into account. In the scheme, the computation and communication procedure form a feedback loop and could not be dissociated. Next, presented a discrete-time two-phase tandem queueing technique. Dai et al. [14] present reconfigurable intelligent surface (RIS) into edge computing for supporting lower-latency applications, in which edge computing could improve the heavier computational pressure of mobile devices with universally distributed computational resources, and RIS improves the quality of wireless transmission connection by perceptively changing the radio propagation environments.

This study develops a novel data offloading with deep learning enabled cyberattack detection (DADL-CAD) model for edge computing in 6G networks. The proposed DADL-CAD technique primarily designs recurrent neural network (RNN) model for traffic flow forecasting in the edge computing enabled 6G networks. Also, adaptive sampling cross entropy (ASCE) model was utilized for maximizing the network efficiency by proper decision making related to the offloading process. Moreover, competitive swarm optimization (CSO) with stacked autoencoder (SAE) technique was executed for the detection of cyberattacks in the network. The performance validation of the DADL-CAD technique is examined in various aspects.

2.2 The Proposed Model

In this study, a new DADL-CAD approach was developed for data offloading and cyberattack detection in 6G networks. The proposed DADL-CAD technique encompasses a series of processes namely RNN based traffic forecasting, ASCE based data offloading, SAE based cyberattack detection, and CSO based parameter tuning. The ASCE model is utilized for maximizing the network efficiency by proper decision making related to the offloading process.

2.2.1 RNN Based Traffic Flow Forecasting

At the initial stage, the RNN model can be employed to forecast the traffic flow [15]. The notation for the single examination case further explicitly equivalents to a series of length.

$$h_1 = \tanh\left(W_{xh}x_1 + b_h\right), \tag{2.1}$$

$$p(\gamma_1 \mid x_1) = \sigma\left(W_{hy}h_1 + b_y\right). \tag{2.2}$$

Then, assume that change in linear transformation utilized from the hidden state calculation was dependent upon not only our input x apart from data in the past that is supported with the hidden state. It has set our preceding hidden state to 0 that is understood as supporting no data in the past:

$$h_0 = 0, \tag{2.3}$$

$$h_1 = \tanh\left(W_{hh}h_0 + W_{xh}x_1 + b_h\right), \tag{2.4}$$

$$p(\gamma_1 \mid x_1) = \sigma\left(W_{hy}h_1 + b_y\right). \tag{2.5}$$

Noticeably, in the modeling perspective, this network was specifically equal to novel network; it varies only from notation and function counts (presented by the unnecessary matrix-vector multiply containing $h_0 = 0$). But, the network is obviously extended to patients with some amount of examinations:

$$h_0 = 0, \tag{2.6}$$

$$h_1 = \tanh\left(W_{hh}h_0 + W_{xh}x_1 + b_h\right), \tag{2.7}$$

$$h_2 = \tanh\left(W_{hh}h_1 + W_{xh}x_2 + b_h\right), \tag{2.8}$$

$$h_T = \tanh\left(W_{hh}h_{T-1} + W_{xh}x_T + b_h\right), \tag{2.9}$$

$$p(\gamma_T \mid x_1, ..., x_T) = \sigma\left(W_{hy}h_T + b_y\right). \tag{2.10}$$

The RNN was able to process series of some lengths, i.e. one with length of $T = 1$ and other with length of $T = 7$, as the transition operation and their parameters are shared over time. If determined, training the network was implemented from the approach that is closely similar to the procedure explained to feedforward network. All the patients are equivalents to an order of exams, $x_1, ..., x_T$, together with label γ_T. It can be procedure for calculating graph with unrolling the RNN on these time steps, and with adding functions for computing the probability of malignances and the loss. Afterward, gradient can be attained using back propagation (BP) and optimizing the utilization of stochastic gradient descent (SGD). The initial RNN variations are established from the literature of easy RNNs or Elman RNN. In analyses, it is classically understanding the further compact representation:

$$h_t = \tanh\left(W_{hh}h_{t-1} + W_{xh}x_t + b_h\right) \tag{2.11}$$

whereas the primary hidden state was omitted, and, unless then identified, is frequently considered that $h_0 = 0$. While the simplified computation graph is equivalent to Eq. (2.11).

2.2.2 ASCE Based Data Offloading

The ASCE model is utilized for maximizing the network efficiency by proper decision making related to the offloading process [16]. It is a most binary integer programming problem that is resolved effectually with branch-and-bound (BnB) technique with an enormous processing difficulty, particularly, when X is superior. In addition, a massive amount of operations are improving the computational access points (CAPs) approach. Then, BnB technologies meet the demand of practical functions. Afterward, the work tries to solve the problem by implementing typical optimized techniques. A notable outcome is for applying convex relaxation like relax $x_{nm} \in 0, 1\}$ as $x_{nm} \in [0, 1]$ by linear programming relaxation (LPr):

$$T(X) = \max_{m \in \mathcal{M}} T_m(X) \, as \, T(X) \geq \max_{m \in \mathcal{M}} T_m(X) \tag{2.12}$$

But relaxation inclines for degrading the efficiency if related to BnB technique. After that, the discrete optimized parameter was solved by executing a probabilistic approach from learning the probabilities of policy x_{nm}. For resolving this issue, cross entropy (CE) technology was executed together with adaptive sampling as ASCE.

2.2.3 SAE Based Cyberattack Detection

Autoencoder is a type of unsupervised learning model that comprises output, input, and hidden layers. The procedure of autoencoders (AE) training comprises encoding and decoding [17]. An encoding was utilized to map the input dataset into hidden depiction, and decoder is represented to recreate input dataset in the hidden depiction. Assume the unlabeled input data $\{x_n\}_{n=1}^{N}$, whereas $x_n \in R^{m \times 1}$, h_n signifies the hidden encoding vector computing from x_n, and $\hat{x}_n$ denotes the decoding vector of the resultant layer:

$$h_n = f(W_1 x_n + b_1) \tag{2.13}$$

In which f represent the encoder function, W_1 indicates the weight matrix, and b_1 shows the bias vector:

$$\hat{x}_n = g(W_2 h_n + b_2) \tag{2.14}$$

Here g indicates the decoder function, W_2 shows the weight matrix, and b_2 indicates the bias vector. The variable set of AEs are enhanced for minimizing the reconstruction error:

$$\emptyset(\Theta) = \arg\min_{\theta, \theta'} \frac{1}{n} \sum_{i=1}^{n} L(x^i, \hat{x}^i) \tag{2.15}$$

While L denotes a loss function $L(x, \hat{x}) = \|x - \hat{x}\|^2$, the architecture of SAE is stacking n autoencoder as to n hidden layers by unsupervised layer-wise learning model and fine-tuned by a supervised model. Therefore, the SAE based method is classified as follows:

1) Training the initial AE by input data and attaining the learned feature vector.

2) The feature vector of previous layer was utilized as the input for the following layer, and this process is repetitive till the training finishes.
3) Afterward each hidden layer is trained, and BP method is utilized for minimizing the cost function and upgrading the weight with trained labeled set to accomplish fine-tuning.

2.2.4 CSO Based Parameter Optimization

To tune the weight and bias values of SAE technique, the CSO algorithm has been applied. CSO is the current swarm intelligence based approach, stimulated from particle swarm optimization (PSO) approach, but the concept is quite distinct from standard PSO [18]. Consider n amount of particles, $S(t)$ represents swarm initialization. All the particles denote a possible solution. The swarm $S(t)$ has n particles, in all the iterations $n/2$ pairs are allocated arbitrarily, and afterward, a competition is made among two particles in all the pairs of particles. Because of competition, the particles have optimal fitness value, named as "winner," and would be directly passed to the following iteration of the swarm, $S(t+1)$, whereas the particle that loses the competition named as a "loser," would upgrade its position and velocity by learning from the winner. In all the iterations, a particle includes once in a competition. Consider the position and velocity of the winner and loser in mth round of competition in iteration t with $V_{w,m}(t)$, $V_{l,m}(t)$, and $X_{w,m}(t)$, $X_{l,m}(t)$ correspondingly, whereas $m = 1, 2, 3, ..., n/2$.

As per the fundamental concept of CSO, afterward mth round of competition the position and velocity of the loser would be upgraded as follows:

$$V_{l,m}(t+1) = r_1(m, t) \times V_{l,m}(t) + r_2(m, t) \times (X_{w,m}(t) - X_{l,m}(t)) + \varphi \times r_3(m, t) \times (\bar{X} - X_{l,m}(t))$$

$$(2.16)$$

$$X_{l,m}(t+1) = X_{l,m}(t) + V_{l,m}(t+1) \qquad (2.17)$$

In which, $r_1(m, t)$, $r_2(m, t)$, and $r_3(m, t) \in [0, 1]$ and $\bar{X}(t)$ represent the mean location and is determined in two manners, global mean and local mean represented by $\bar{X}_m^g(t) -$ and $X_l^g(t)$. $X_m^g(t)$ signifies the global mean location of each particle, where $\bar{X}_l^g(t)$ – represents local mean of predetermined neighborhood of particle k. φ denotes the variables that control the impact of $\bar{X}_m(t)$.

To understand better about the CSO, the study presented related comparison using PSO:

1) The initial part $r_1(m, t) \times V_{l, m}(t)$ is analogous to inertia term in PSO, that balance the motion of swarm, the variance is inertia weight w, which is exchanged as an arbitrary vector $r_1(m, t)$ in CSO.

2) The next part $r_2(m, t) \times (X_{w, m}(t) - X_{l, m}(t))$ is analogous to the cognitive element in PSO, but it is theoretically quite distinct in PSO, particle loses its competition learn from winner, rather than personal best found until now.

   This method is more feasible biologically when inspiring swarm behavior since it is difficult for remembering own best experience.

3) The last part $r_3(m, t) \times (\bar{X}(t) - X_{l,m}(t))$ is analogous to the social element in standard PSO, but particles lose competition learned from the mean location rather *gbest* in PSO; no memory is needed that might biologically be very feasible.

The feed forward (FF) was utilized for authenticating the quality of particles. The objective of CSO is to discover the particle position that results in optimal assessment of fitness function. All the particles are allocated to an arbitrary velocity and position in the initialization method of CSO. PSO has set the standard in different applications including optimum feature selection problems in software costing, machine learning algorithm, routing, clustering, and so on. CSO attained the optimal result in the tested standard function than PSO, hence, it motivates us to employ in cluster head (CH) selection and routing problem as stated in the presented method.

2.3 Performance Validation

In this section, the result analysis of the DADL-CAD technique with respect to different aspects. First, the normalized mean square error (NMSE) analysis of DADL-CAD algorithm with recent DL algorithms takes place in Table 2.1 and Figure 2.3. The outcomes indicated that the DADL-CAD algorithm has resulted in lowering values of NMSE under distinct areas. For instance, with area of $5\,km^2$, the DADL-CAD algorithm has offered lower NSME of 0.009 43 while the LSTM, BiLSTM, and GRU algorithms have reached maximum NMSE of 0.001 348, 0.011 05, and 0.009 81, respectively. In addition, with area of $10\,km^2$, the DADL-CAD algorithm has offered lower NMSE of 0.006 15, while the LSTM, BiLSTM, and GRU algorithms have reached maximum NMSE of 0.006 15, 0.004 49, and 0.003 20, respectively. Moreover, with area of $15\,km^2$, the DADL-CAD algorithm has provided minimum NMSE of 0.004 16, while the LSTM, BiLSTM, and GRU algorithms have gained higher NMSE of 0.002 75, 0.002 18, and 0.001 97 correspondingly.

Furthermore, with area of $20\,km^2$, the DADL-CAD algorithm has offered lower normalized mean moderate error (NMME) of 0.001 99 while the LSTM, BiLSTM, and GRU algorithms have reached maximum NMSE of 0.000 82, 0.000 62, and 0.000 33, respectively.

Table 2.1 Traffic Prediction analysis of DADL-CAD model with respect to NMSE.

Area (km^2)	LSTM	BiLSTM	GRU model	Proposed model
5	0.013 48	0.011 05	0.009 81	0.009 43
10	0.006 15	0.004 49	0.003 20	0.002 72
15	0.004 16	0.002 75	0.002 18	0.001 97
20	0.003 60	0.001 86	0.001 59	0.001 08
25	0.001 99	0.000 86	0.000 62	0.000 33
30	0.000 84	0.000 37	0.000 27	0.000 12
35	0.000 78	0.000 23	0.000 10	0.000 08
40	0.000 73	0.000 27	0.000 10	0.000 05
Average	**0.003 97**	**0.002 74**	**0.002 23**	**0.001 97**

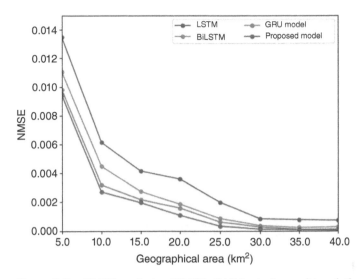

Figure 2.3 NMSE analysis of DADL-CAD technique with existing approaches.

Figure 2.4 demonstrates the average NSME analysis of the DADL-CAD model with recent methods. The results indicated that the DADL-CAD model has accomplished effectual outcomes with minimal average NMSE of 0.003 97, but the LSTM, BiLSTM, and GRU techniques have accomplished higher average NMSE of 0.002 74, 0.002 23, and 0.001 97 correspondingly.

Figure 2.5 depicts the overall task completion time analysis of the DADL-CAD method. The results indicated that the ASCE technique has obtained lower task completion times compared to adaptive and fixed models. In addition, the ASCE model has offered a lower average task completion time of 0.680 while the adaptive and fixed models have obtained higher average task completion times of 1.718 and 1.867 correspondingly.

The cyberattack detection performance of the DADL-CAD method is examined with recent methods on distinct sizes of training and testing data. The outcomes indicated that the DADL-CAD model has resulted in effective classification outcome as illustrated in Table 2.2 [16].

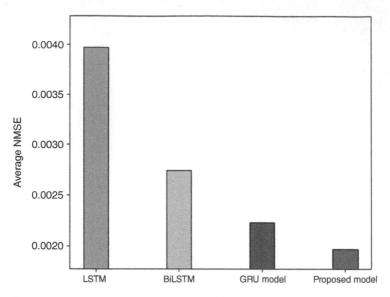

Figure 2.4 Average NMSE analysis of DADL-CAD technique with existing approaches.

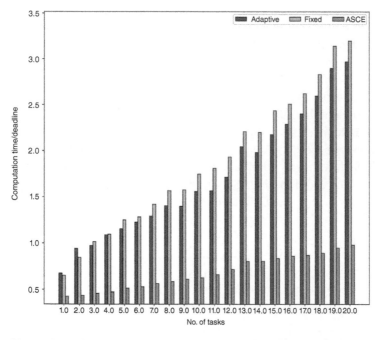

Figure 2.5 Computation time analysis of DADL-CAD technique.

Table 2.2 Performance evaluation of existing with proposed technique on attack detection dataset.

Methods	$Sens_y$	$Spec_y$	Acc_y	F_{score}	Kappa
Training size = 60% testing size = 40%					
Proposed model	99.60	98.20	98.90	99.10	97.20
O-FFNN	99.40	97.50	98.10	98.60	96.90
O-DBN	99.00	96.20	97.70	97.80	95.30
DBN model	91.50	95.80	96.20	91.60	91.30
Random forest	92.40	93.80	93.00	93.60	86.00
Training size = 70% testing size = 30%					
Proposed model	99.70	98.90	99.20	99.40	98.60
O-FFNN	99.60	97.90	98.90	99.10	97.40
O-DBN	99.30	96.60	98.20	98.00	96.10
DBN model	92.50	96.30	96.80	92.60	92.70
Random forest	93.80	94.50	93.80	94.40	89.20

Figure 2.6 demonstrates the cyberattack detection results of the DADL-CAD model with other methods on training/testing data of 60 : 40. The results indicated that the DADL-CAD model has resulted in effectual outcome in terms of different measures. With respect to $sens_y$, the DADL-CAD model has gained improved $sens_y$ of 99.60% whereas the O-FFNN, O-DBN, DBN, and RF models have obtained lower $sens_y$ of 99.40, 99, 91.50, and 92.40%, respectively. Moreover, with respect to acc_y, the DADL-CAD model has gained improved acc_y of 98.90% whereas the O-FFNN, O-DBN, DBN, and RF models have obtained lower acc_y of 98.10, 97.70, 96.20, and 93% correspondingly. Furthermore, with respect to F_{score}, the DADL-CAD algorithm has gained improved F_{score} of 99.10% whereas the O-FFNN, O-DBN, DBN, and RF algorithms have obtained lower F_{score} of 98.60, 97.80, 91.60, and 93.60% correspondingly.

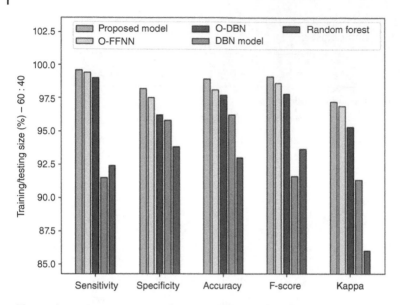

Figure 2.6 Comparative analysis of DADL-CAD technique under training/ testing (60 : 40) dataset.

Figure 2.7 illustrates the cyberattack detection results of the DADL-CAD approach with other methods on training/testing data of 70 : 30. The outcomes exposed that the DADL-CAD model has resulted in effectual outcomes in terms of different measures. With respect to $sens_y$, the DADL-CAD algorithm has gained improved $sens_y$ of 99.70% whereas the O-FFNN, O-DBN, DBN, and RF models have obtained lower $sens_y$ of 99.60, 99.30, 92.50, and 93.80%, respectively. Also, with respect to acc_y, the DADL-CAD technique has gained improved acc_y of 99.20% whereas the O-FFNN, O-DBN, DBN, and RF systems have obtained lower acc_y of 98.90, 98.20, 96.80, and 93.80%, respectively. At last, with respect to F_{score}, the DADL-CAD technique has gained higher F_{score} of 99.40% whereas the O-FFNN, O-DBN, DBN, and RF techniques have reached lower F_{score} of 99.10, 98, 92.60, and 94.40% correspondingly.

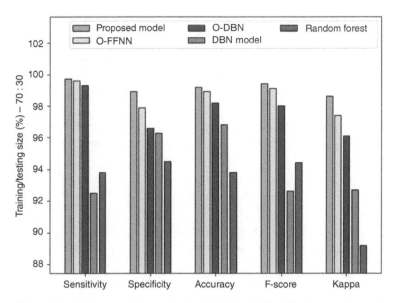

Figure 2.7 Comparative analysis of DADL-CAD technique under training/testing (70 : 30) dataset.

2.4 Conclusion

In this study, a new DADL-CAD technique has been developed for data offloading and cyberattack detection in 6G networks. The proposed DADL-CAD technique encompasses a series of processes namely RNN based traffic forecasting, ASCE based data offloading, SAE based cyberattack detection, and CSO based parameter tuning. The ASCE model is utilized for maximizing the network efficiency by proper decision making related to the offloading process. The performance validation of the DADL-CAD technique is examined under various aspects, and the comparative study reported the supremacy of the DADL-CAD technique over the recent approaches. Therefore, the DADL-CAD technique can be considered as an effectual tool for improving the efficiency of the MEC enabled 6G networks. In future, data aggregation models can be developed to reduce communication process in 6G networks.

References

1 Lovén, L., Leppänen, T., Peltonen, E. et al. (2019). EdgeAI: a vision for distributed, edge-native artificial intelligence in future 6G networks. *Proceedings of the First 6G Wireless Summit*, pp. 1–2, California, USA (23-Aug 2019).

2 Al-Ansi, A., Al-Ansi, A.M., Muthanna, A. et al. (2021). Survey on intelligence edge computing in 6G: characteristics, challenges, potential use cases, and market drivers. *Future Internet* 13 (5): 118.

3 Tomkos, I., Klonidis, D., Pikasis, E., and Theodoridis, S. (2020). Toward the 6G network era: opportunities and challenges. *IT Professional* 22 (1): 34–38.

4 Ergen, M., Inan, F., Ergen, O. et al. (2020). Edge on wheels with OMNIBUS networking for 6G technology. *IEEE Access* 8: 215928–215942.

5 Ishtiaq, M., Saeed, N., and Khan, M.A. (2021). Edge computing in IoT: A 6G perspective. *arXiv* 23: 23–34.

6 Hui, Y., Cheng, N., Huang, Y. et al. (2021). Personalized vehicular edge computing in 6G. *IEEE Network* 35 (6): 278–284.

7 Xiao, Y., Shi, G., Li, Y. et al. (2020). Toward self-learning edge intelligence in 6G. *IEEE Communications Magazine* 58 (12): 34–40.

8 Zhao, L., Zhou, G., Zheng, G. et al. (2021). Open-source-defined multi-access edge computing for 6G: opportunities and challenges. *IEEE Access* 9: 158426–158439.

9 Huynh, L.N. and Huh, E.N. (2021). Envisioning edge computing in future 6G wireless networks. *2021 Fifth World Conference on Smart Trends in Systems Security and Sustainability (WorldS4)*, pp. 307–311, London, UK (29–30 July 2021). UK: IEEE.

10 Koketsurodrigues, T., Liu, J., and Kato, N. (2021). Offloading decision for mobile multi-access edge computing in a multi-tiered 6G network. *IEEE Transactions on Emerging Topics in Computing* 24: 1–10.

11 Rodrigues, T.K., Liu, J., and Kato, N. (2021). Application of cybertwin for offloading in mobile multiaccess edge computing for 6G networks. *IEEE Internet of Things Journal* 8 (22): 16231–16242.

12 Liao, Z., Peng, J., Huang, J. et al. (2020). Distributed probabilistic offloading in edge computing for 6G-enabled massive Internet of Things. *IEEE Internet of Things Journal* 8 (7): 5298–5308.

13 Cao, J., Feng, W., Ge, N., and Lu, J. (2020). Delay characterization of mobile-edge computing for 6G time-sensitive services. *IEEE Internet of Things Journal* 8 (5): 3758–3773.

14 Dai, Y., Guan, Y.L., Leung, K.K., and Zhang, Y. (2021). Reconfigurable intelligent surface for low-latency edge computing in 6G. *IEEE Wireless Communications* 28 (6): 72–79.

15 Medsker, L.R. and Jain, L.C. (2001). Recurrent neural networks. *Design and Applications* 5: 64–67.

16 Gopalakrishnan, T., Ruby, D., Al-Turjman, F. et al. (2020). Deep learning enabled data offloading with cyber attack detection model in mobile edge computing systems. *IEEE Access* 8: 185938–185949.

17 Zhou, P., Han, J., Cheng, G., and Zhang, B. (2019). Learning compact and discriminative stacked autoencoder for hyperspectral image classification. *IEEE Transactions on Geoscience and Remote Sensing* 57 (7): 4823–4833.

18 Cheng, R. and Jin, Y. (2014). A competitive swarm optimizer for large scale optimization. *IEEE Transactions on Cybernetics* 45 (2): 191–204.

3

Henry Gas Solubility Optimization with Deep Learning Enabled Traffic Flow Forecasting in 6G Enabled Vehicular Networks

José Escorcia-Gutierrez[1,2], Melitsa Torres-Torres[3], Natasha Madera[4], and Carlos Soto[5]

[1] Research Center - CIENS, Escuela Naval de Suboficiales ARC Barranquilla, Barranquilla, Colombia
[2] Biomedical Engineering Program, Corporación Universitaria Reformada, Barranquilla, Colombia
[3] Research Group IET-UAC, Universidad Autónoma del Caribe, Barranquilla, Colombia
[4] Mechatronics Engineering Program, Universidad Simón Bolívar, Barranquilla, Colombia
[5] Mechanical Engineering Program, Universidad Autónoma del Caribe, Barranquilla, Colombia

3.1 Introduction

The constant progression to the 6G wireless transmission technique overcomes stringent computation, storage, power, and privacy limitations for making an intelligent and efficient next generation transport scheme to improve driving experience and mitigate traffic jams in vehicular ad hoc networks (VANETs) [1, 2]. Along with 6G technique, higher throughput, higher availability, and higher reliability are empowered in VANET. Due to the infrastructure improvement and economic growth, work and people's lives have not been confined to a certain city. The social activity related to many cities became an essential prerequisite of our day-to-day life [3]. Like a connection between two cities, highways certainly play a major role in our day-to-day life and work. When a road is occasionally closed or traffic jam occurs because of bad weather conditions or an accident on the highways, correspondingly, people's work or travel plans would be affected severely [4]. Like a traffic

AI-Enabled 6G Networks and Applications, First Edition. Edited by Deepak Gupta, Mahmoud Ragab, Romany Fouad Mansour, Aditya Khamparia, and Ashish Khanna.
© 2023 John Wiley & Sons Ltd. Published 2023 by John Wiley & Sons Ltd.

supervision method, traffic flow predictive technique for highways could assist the relevant departments and government in dispatching vehicles to avoid traffic congestion and for making road planning [5]. Since highway traffic predictive method has spatial and temporal features as well influenced by the external factor, inadequate consideration of factor influencing is the major problem confronted by the traffic flow prediction of highway toll stations. Figure 3.1 illustrates the infrastructure of VANET.

The grouping of spatial and temporal could be named as spatiotemporal features. External factor includes date type and weather condition. For instance, people are better prepared for taking a road trip in good weather. Particularly on weekends or holidays, the demand for folks enjoying travel increases sharply. Hence, to efficiently utilize this factor on past information has become the emphasis of highway traffic flow

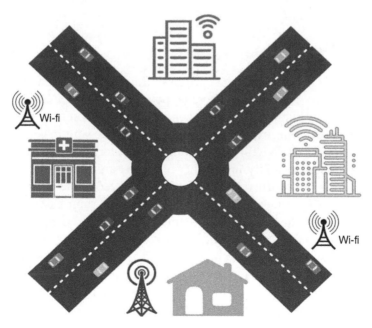

Figure 3.1 VANET structure.

prediction [6]. Indeed, short-term traffic flow prediction relies mostly on the real-time and historical traffic information gathered by different sensor nodes (for example, radar, induction coils, social media, cameras, and mobile global positioning systems) for building respective algorithms and models [7]. Through principles, present short-term traffic flow predictive models are widely categorized into simulation approaches, parametric methods, and nonparametric methods [7].

Bhatia et al. [8] presented a data-driven method to implement an AI method for vehicular traffic behavior forecasting. Then they integrated the adaptability, flexibility, and scalability leveraged by the software defined vehicular network (SDVN) framework and machine learning (ML) approaches for efficiently modeling the traffic flow. Then constructed an long short term memory neural network (LSTM-NN) framework that overcomes the topic of backpropagated error decay via memory block for spatiotemporal traffic predictive model using higher temporal dependency. Zhao et al. [9] presented a traffic accident predictive method based convolution neural network (CNN). The DL approach could extract autonomous features from a massive number of information gathered in VANET. Distinct convolutional kernels are accountable to extract distinct characteristics, and attained novel parameters are input into the trained model from the edge computing server for testing and training [10, 11]. Ref. [12] proposed a centralized routing system using mobility forecast to VANET aided by AI-driven software-determined network (SDN). Especially, the SDN controller could implement precise mobility predictive method by using an innovative artificial neural network (ANN) method. Next, depending on the mobility predictive system, the effective communication possibility and normal delay of every vehicle request under common network topology variations are evaluated. Tong et al. [13] employ an optimization of support vector regression (SVM) as a method of traffic flow prediction. Particle swarm optimization has been employed for the parameter optimized of support vector regressor (SVR) that enhances the efficiency of the predictive method.

This study develops a novel Henry gas solubility optimization with deep learning enabled traffic flow forecasting (HGSODL-TFF) technique for 6G enabled vehicular networks. The presented HGSODL-TFF technique primarily intends to forecast the level of traffic in the 6G enabled VANET. In addition, the HGSODL-TFF model initially preprocesses the traffic data using z-score normalization approach. Besides, deep belief network (DBN) model is employed to effectually forecast the traffic flow. Also, HSGO algorithm can be applied for optimally modifying the hyperparameters (such as learning rate, epoch count, and batch size) of the DBN model thereby improving the forecasting performance. The experimental validation of the HGSODL-TFF model is performed on test data and the results are inspected under several aspects.

3.2 The Proposed Model

In this study, a new HGSODL-TFF technique has been developed to predict traffic flow in the 6G assisted VANET. The presented HGSODL-TFF model encompasses three distinct processes such as preprocessing, DBN based prediction, and HSGO based hyperparameter tuning. The details involved in each module are discussed in the following sections.

3.2.1 Z-Score Normalization

The z-score is a convention standardized and normalized approach that signifies the amount of standard deviation (SD), a raw datapoint occurring above/below the population mean. It perfectly lies among -3 and $+3$. It normalizes the dataset to aforementioned scale for converting each information with different scales to default scales. For normalizing the information utilizing the z-score, it can be subtracted from the mean of population in a raw datapoint and separated by SD that provides a score ideally different among -3 and $+3$, so, the reflecting several SDs are a point above or below the mean as calculated as in Eq. (3.1), whereas x implies the value of specific instance, μ denotes the mean, and σ stands for the SD.

$$z - \text{score} = \frac{(x - \mu)}{\sigma} \tag{3.1}$$

3.2.2 DBN Based Prediction Model

Once the input data is preprocessed, the DBN model is employed for the forecasting of traffic flow data. DBN is a building block of DNN that encompasses distinct layers such as multilayer perceptron (MLP) and restricted Boltzmann machine (RBM) [14, 15]. RBM consists of hidden unit and visible unit that are interconnected according to the weighted connection. MLP is adopted as feed forward neural network (FFNN) that contains hidden, input, and output layers. Let us assume there are two RBMs, such as RBM1 and RBM2, and the input to RBM1 is the feature vector attained from the big data. Figure 3.2 demonstrates

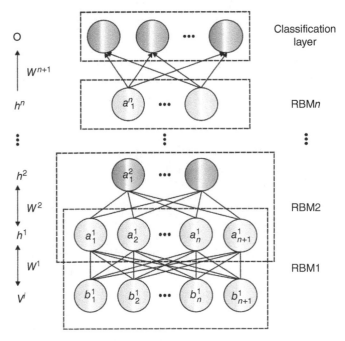

Figure 3.2 Structure of DBN.

the framework of DBN technique. The hidden and input layers in the input neuron of RBM1 is given as follows:

$$\varepsilon^1 = \left\{\varepsilon_1^1, \varepsilon_2^1, \varepsilon_3^1, ..., \varepsilon_i^1, ..., \varepsilon_m^1\right\}; \quad 1 \le i \le m \tag{3.2}$$

$$\beta^1 = \left\{\beta_1^1, \beta_2^1, ..., \beta_c^1, ..., \beta_f^1\right\}; \quad 1 \le c \le f \tag{3.3}$$

Where ε_i^1 signifies the ith input layer that is existing in RBM1 and the amount of input layers of RBM1 was equivalent to the dimensional feature vector. There is m neuron from the input layer of RBM1 for implementing classification. Consider the overall amount of the hidden layers in the RBM1 be f and cth hidden layer from RBM2 be β_c^1. The bias of hidden and visible layers of RBM1 are given as follows:

$$A^1 = \left\{A_1^1, A_2^1, A_3^1, ..., A_i^1, ..., A_m^1\right\} \tag{3.4}$$

$$O^1 = \left\{O_1^1, O_2^1, ..., O_c^1, ..., O_f^1\right\} \tag{3.5}$$

The bias of input and hidden layer of RBM1 is equivalent to the overall neurons in the weights and layers of RBM1 are shown in the following:

$$a^1 = \left\{a_{ic}^1\right\}; \quad 1 \le i \le m; \quad 1 \le c \le f \tag{3.6}$$

Where a_{ic}^1 characterizes the weight of RBM1, and it can be weight among ith input layer and cth hidden layer of RBM1. Henceforth, the resultant of RBMI can be shown as follows:

$$\beta_c^1 = fun\left[O_c^1 + \sum_i \tau_i^1 a_{ic}^1\right] \tag{3.7}$$

Here fun characterizes the activation function from RBM1 and τ_i^1 signifies the feature vector as. The output of RBM1 can be formulated by,

$$\beta^1 = \{\beta_c^1\}; \quad 1 \le c \le f \tag{3.8}$$

The output from RBM1 is given as the input to RBM2 and the resultant of RBM2 is assessed by the preceding equation. The resultant of

RBM2 is shown as G_j^2, that is provided as the input to the MLP layer. The input neuron in MLP is formulated by,

$$\lambda = \{\lambda_1, \lambda_2, ..., \lambda_e, ..., \lambda_f\}; \quad 1 \leq c \leq f \tag{3.9}$$

While f shows the overall input layers from the MLP. The hidden neuron of MLP is shown as,

$$E = \{E_1, E_2, ..., E_x, ..., E_y\}; \quad 1 \leq x \leq y \tag{3.10}$$

In the equation, y shows the overall hidden layers of MLP. The bias of hidden neuron is formulated by,

$$B = \{B_1, B_2, ..., B_w, ..., B_z\}; \quad 1 \leq w \leq z \tag{3.11}$$

Where z signifies the output neuron in the MLP layer. The weights among the input and hidden neurons are shown as,

$$a^{\text{mlp}} = \{a_{cx}^{\text{mlp}}\}; 1 \leq c \leq f; 1 \leq x \leq y \tag{3.12}$$

Let u_{cx}^{mlp} be the weight vectors among cth input layer and the xth hidden layer. The resultant of hidden neurons in MLP is depending on the weights and bias,

$$O^{hid} = \left[\sum\nolimits_{c\,=\,1}^{f} a_{cx}^{\text{mlp}} \times F_c\right] a_x \tag{3.13}$$

Hence a_x shows the bias of output neurons. The weight vectors among the output and hidden layers are shown as follows:

$$a' = \{a'_{xw}\}; 1 \leq x \leq y; \quad 1 \leq w \leq z \tag{3.14}$$

Here, the output of MLP can be evaluated by,

$$B_w = \sum\nolimits_{x\,=\,1}^{y} a'_{xw} \times O^{hid} \tag{3.15}$$

In which a'_{xw} shows weights among the output and hidden layers in MLP, and O^{hid} denotes the output of the hidden neuron.

3.2.3 HSGO Based Hyperparameter Optimization Model

For optimally tuning the hyperparameters (such as learning rate, epoch count, and batch size) of the DBN model, the HSGO algorithm is applied. The HGSO is a new multi hybrid (MH) technique that inspires Henry's law [16]. Compared to population based techniques, the HGSO begins with primary value to a group of N gases or solutions (S) and this is dependent upon the searching space and is expressed as:

$$X_i = Lb + rand * (Ub - Lb), \quad rand \in [O, 1] \tag{3.16}$$

where Lb and Ub stand for the lower as well as upper values from the searching space correspondingly. The gas set X is clustered as to equivalent amount of groups N_g, and all the groups signify a kind of gas. All the groups have a similar value to Henry's constant that is determined as:

$$H_j = l \times r_1, \quad j = 1, 2, ..., N_g, \quad l = 5E - 2 \tag{3.17}$$

where r_1 denotes the arbitrary number and l indicates the constants [17]. The next step is for determining optimum gas in all the groups. Afterward, entire cluster groups of optimum gases are defined. Afterward, Henry's coefficient (H_j) to the jth group was upgraded as:

$$H_j(t+1) = H_j(t) \times \exp\left(-C_j \times \left(\frac{1}{T(t)} - \frac{1}{T^\theta}\right)\right), \quad T(t) = \exp\left(-\frac{t}{iter}\right) \tag{3.18}$$

where, T^θ, and $iter$ represents the temperature, constant value (fixed to 298.15), and the maximal amount of iterations correspondingly. The HGSO upgrades the solubilities (S_{ij}) of X_j, $(i = 1, 2, ..., N)$ among every group as:

$$S_{ij}(t) = K \times H_j(t+1) \times P_{ij}(t) \tag{3.19}$$

whereas K refers the constant and $P_j(t)$ implies the partial pressure on ith gas from the jth cluster j, and it can be demonstrated as:

$$P_{ij}(t) = l_2 \times r_1, \quad j = 1, 2, ..., N_g, \quad l_2 = 100 \tag{3.20}$$

The gas X_i that goes to the jth cluster was upgraded based on Eq. (3.21).

$$X_{ij}(t+1) = X_{ij}(t) + F_g \times r \times \eta \times \left(X_{ib}(t) - X_{ij}(t)\right)$$
$$+ F_g \times r \times \alpha \times \left(S_{ij}(t) \times X_{ib}(t) - X_{ij}(t)\right) \tag{3.21}$$

where $\eta = \beta \times \exp\left(-\dfrac{F_b(t)+e}{F_{ij}(t)+e}\right)$ that determined the capability of ith gases for interacting with other gases from jth group. F_{ij} implies the fitness value of X_j on jth group and F_b signifies the optimum fitness value. F_g refers the fag value that modifies the direction of gases. The $\alpha = l$ denotes the control of other gases on X_j from the group j. For escaping the local points, the HGSO upgrades the worse N_w gases as in Eq. (3.22).

$$G_{ij} = G_{ij}^{\min} + r \times \left(G_{ij}^{\max} - G_{ij}^{\min}\right), \quad i = 1, 2, ..., N_w \tag{3.22}$$

$$N_w = N \times r \times (c_2 - c_1) + c_1, \quad c_1 = 0.1, \quad \text{and} \quad c_2 = 0.2 \tag{3.23}$$

G_{ij} defines the gas X_i from jth group that goes to the worse gas.

3.3 Experimental Validation

The performance validation of the HGSODL-TFF model is carried out using the traffic data collected by our own in the 6G enabled vehicular networks. The forecasting results are inspected under different test runs and durations. Table 3.1 and Figure 3.3 provide the actual and predicted traffic volume of the HGSODL-TFF model under five distinct runs. The results indicated that the HGSODL-TFF model has resulted in maximum prediction performance under all days and runs. For instance, on day 1, the HGSODL-TFF model has attained predicted values of 24 257, 24 178, 23 832, 24 124, and 24 206 under runs 1–5, respectively.

Similarly, on day 4, the HGSODL-TFF approach has reached predicted values of 24 311, 24 896, 24 736, 24 790, and 24 729 under runs 1–5, respectively. Likewise, on day 6, the HGSODL-TFF system has attained predicted values of 27 504, 25 641, 26 732, 26 652, and 27 064

Table 3.1 Predicted traffic volume analysis of HGSODL-TFF technique under five distinct runs.

No. of days	Ground truth	Run-1	Run-2	Run-3	Run-4	Run-5
1	24 071	24 257	24 178	23 832	24 124	24 206
2	24 869	25 242	24 763	24 976	25 003	24 937
3	25 003	24 630	25 162	25 136	24 763	24 882
4	24 630	24 311	24 896	24 736	24 790	24 729
5	24 949	24 577	24 790	25 082	24 869	24 890
6	27 158	27 504	25 641	26 732	26 652	27 064
7	24 444	25 269	24 843	24 630	24 630	24 558
8	24 843	25 455	24 683	25 136	24 923	24 761
9	26 200	26 493	25 375	26 493	25 960	26 102
10	25 082	25 561	24 869	25 242	24 843	25 229

under runs 1–5 correspondingly. Along with that, on day 8, the HGSODL-TFF system has reached predicted values of 25 455, 24 683, 25 136, 24 923, and 24 761 under runs 1–5 correspondingly. Lastly, on day 10, the HGSODL-TFF approach has achieved predicted values of 25 561, 24 869, 25 242, 24 843, and 25 229 under runs 1–5 correspondingly.

A comparative traffic volume forecasting results of the HGSODL-TFF model with existing models is performed in Table 3.2 and Figure 3.4. The results indicated that the HGSODL-TFF model has outperformed the existing techniques under distinct days. For instance, on day 1, the HGSODL-TFF model has effectively predicted the traffic volume as 24 206, whereas the Hyperbolic Graph Convolutional Networks (HGCN), Long Short-Term Memory (LSTM), Gradient Boosted Regression Trees (GBRT), and Nearest Neighbors Algorithm (KNN) models have obtained the traffic volume as 23 885, 22 900, 22 049, and 21 384, respectively.

Simultaneously, on day 4, the HGSODL-TFF method has effectually predicted the traffic volume as 24 729 whereas the HGCN, LSTM, GBRT, and KNN techniques have reached the traffic volume as 25

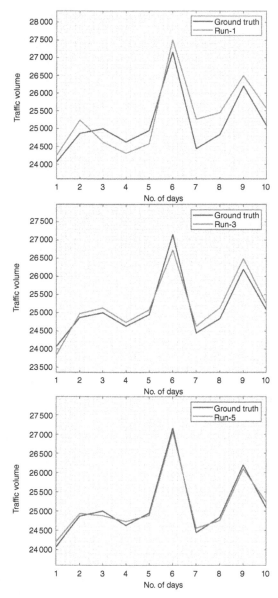

Figure 3.3 Predicted traffic volume analysis of HGSODL-TFF technique under five distinct runs.

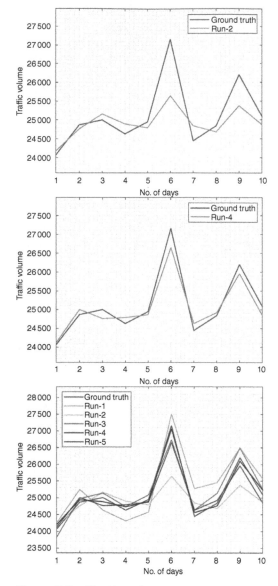

Figure 3.3 (Continued)

Table 3.2 Predicted traffic volume analysis of HGSODL-TFF technique with existing approaches.

No. of days	Ground truth	HGSODL-TFF	HGCN	LSTM	GBRT	KNN
1	24 071	24 206	23 885	22 900	22 049	21 384
2	24 869	24 937	23 672	24 976	22 794	22 049
3	25 003	24 882	26 546	25 269	22 741	22 049
4	24 630	24 729	25 029	23 645	22 315	21 543
5	24 949	24 890	24 497	24 364	22 741	21 996
6	27 158	27 064	25 269	24 444	24 444	23 645
7	24 444	24 558	25 881	25 242	22 235	21 410
8	24 843	24 761	23 167	23 406	22 741	22 155
9	26 200	26 102	25 375	24 524	23 645	22 900
10	25 082	25 229	24 124	24 470	22 847	22 129

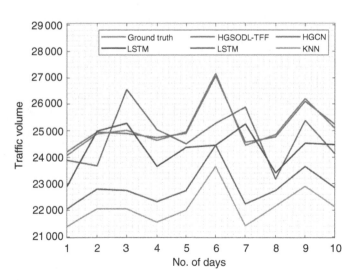

Figure 3.4 Traffic volume analysis of HGSODL-TFF technique with recent methods.

029, 23 645, 22 315, and 21 543 correspondingly. Eventually, on day 6, the HGSODL-TFF approach has efficiently predicted the traffic volume as 24 558 whereas the HGCN, LSTM, GBRT, and KNN algorithms have reached the traffic volume as 25 881, 25 242, 22 235, and 21 410 correspondingly. Meanwhile, on day 8, the HGSODL-TFF system has efficiently predicted the traffic volume as 24 761 whereas the HGCN, LSTM, GBRT, and KNN techniques have obtained the traffic volume as 23 167, 23 406, 22 741, and 22 155, respectively. Lastly, on day 10, the HGSODL-TFF model has effectively predicted the traffic volume as 25 229 whereas the HGCN, LSTM, GBRT, and KNN methodologies have achieved the traffic volume as 24 124, 24 470, 22 847, and 22 129 correspondingly.

In order to demonstrate the betterment of the HGSODL-TFF model, a comparison study with recent methods is provided in Table 3.3 [18]. A comparative root mean square error (RMSE) examination of the HGSODL-TFF model with existing methods takes place in Figure 3.5. The results show that the HGCN model has resulted in poor performance with a maximum RMSE of 1194.72. KNN and LSTM model provide better performance than GBRT and achieve RMSE of 387.86. However, the HGSODL-TFF model has accomplished superior outcomes with the least RMSE of 332.47.

Table 3.3 Comparative analysis of HGSODL-TFF technique with existing approaches in terms of RMSE, mean average error (MAE), and mean absolute percentage error (MAPE).

Methods	RMSE	MAE	MAPE (%)
HGSODL-TFF	332.47	257.54	2.41
HGCN	1194.72	956.93	3.68
GBRT	387.86	346.01	5.47
KNN	387.86	346.01	5.47
LSTM	387.86	346.01	5.47

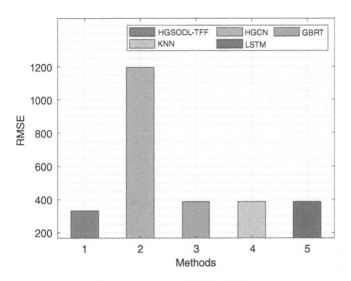

Figure 3.5   RMSE analysis of HGSODL-TFF technique with recent approaches.

A detailed MAE examination of the HGSODL-TFF system with existing methods takes place in Figure 3.6. The outcomes demonstrated that the HGCN technique has resulted in poor performance with a maximum MAE of 956.93. Along with that, the GBRT, KNN, and LSTM models have accomplished slightly improved performance with the identical MAE of 346.01. However, the HGSODL-TFF technique has accomplished maximum outcome with the least MAE of 257.54.

A comparison MAPE examination of the HGSODL-TFF model with existing methods takes place in Figure 3.7. The outcomes demonstrated that the HGCN model has resulted in worse performance with a maximum MAPE of 3.68%. KNN, and LSTM provide better performance than GBRT with MAPE of 5.47%. At last, the HGSODL-TFF methodology has accomplished superior outcomes with the least MAPE of 2.41%.

The aforementioned results indicated the supremacy of the HGSODL-TFF model on the forecasting performance over the other methods.

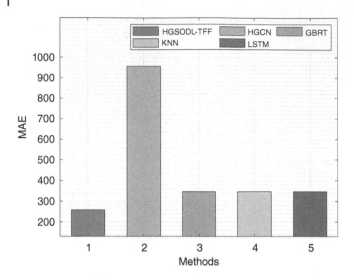

Figure 3.6 MAE analysis of HGSODL-TFF technique with recent approaches.

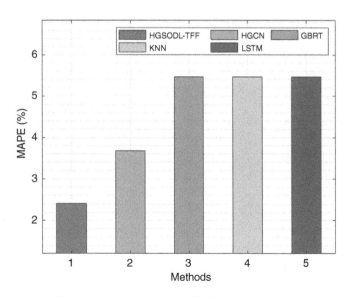

Figure 3.7 MAPE analysis of HGSODL-TFF technique with recent approaches.

3.4 Conclusion

In this study, a new HGSODL-TFF technique has been developed to predict traffic flow in the 6G assisted VANET. The presented HGSODL-TFF model encompasses three distinct processes such as preprocessing, DBN based prediction, and HSGO based hyperparameter tuning. The HSGO algorithm can be applied for optimally modifying the hyperparameters (such as learning rate, epoch count, and batch size) of the DBN model and thereby improving the forecasting performance. The experimental validation of the HGSODL-TFF model is performed on test data, and the results are inspected under several aspects. The simulation results reported the betterment of the HGSODL-TFF model over the other recent approaches. In future, hybrid DL models can be utilized to boost the predictive results of the HGSODL-TFF model.

References

1 Suresh Kumar, K., Radha Mani, A.S., Sundaresan, S., and Ananth Kumar, T. (2021). Modeling of VANET for future generation transportation system through edge/fog/cloud computing powered by 6G. In: *Cloud and IoT-Based Vehicular Ad Hoc Networks* (eds. Gurinder Singh, Vishal Jain, Jyotir Moy Chatterjee et al.), 105–124. Scrivener Publishing.

2 Zhou, Z., Gaurav, A., Gupta, B.B. et al. (2021). A fine-grained access control and security approach for intelligent vehicular transport in 6G communication system. *IEEE Transactions on Intelligent Transportation Systems* 32: 1–10.

3 Vijayakumar, P., Azees, M., Kozlov, S.A., and Rodrigues, J.J. (2021). An anonymous batch authentication and key exchange protocols for 6G enabled VANETs. *IEEE Transactions on Intelligent Transportation Systems.* 32: 1–15.

4 Mchergui, A., Moulahi, T., and Zeadally, S. (2021). Survey on Artificial Intelligence (AI) techniques for Vehicular Ad-hoc Networks (VANETs). *Vehicular Communications* 34: 100403.

5 Tao, Y., Sun, P., and Boukerche, A. (2019). A novel travel-delay aware short-term vehicular traffic flow prediction scheme for VANET. *2019 IEEE Wireless Communications and Networking Conference (WCNC)*, pp. 1–6. IEEE, Marrakech, Morocco (15–19 April 2019).

6 Sun, P., AlJeri, N., and Boukerche, A. (2020). DACON: a novel traffic prediction and data-highway-assisted content delivery protocol for intelligent vehicular networks. *IEEE Transactions on Sustainable Computing* 5 (4): 501–513.

7 Khatri, S., Vachhani, H., Shah, S. et al. (2021). Machine learning models and techniques for VANET based traffic management: implementation issues and challenges. *Peer-to-Peer Networking and Applications* 14 (3): 1778–1805.

8 Bhatia, J., Dave, R., Bhayani, H. et al. (2020). SDN-based real-time urban traffic analysis in VANET environment. *Computer Communications* 149: 162–175.

9 Zhao, H., Cheng, H., Mao, T., and He, C. (2019). Research on traffic accident prediction model based on convolutional neural networks in VANET. *2019 Second International Conference on Artificial Intelligence and Big Data (ICAIBD)*, pp. 79–84, Canada (4 August 2019). Canada: IEEE.

10 Abdelatif, S., Derdour, M., Ghoualmi-Zine, N., and Marzak, B. (2020). VANET: a novel service for predicting and disseminating vehicle traffic information. *International Journal of Communication Systems* 33 (6): e4288.

11 Nadarajan, J. and Kaliyaperumal, J. (2021). QOS aware and secured routing algorithm using machine intelligence in next generation VANET. *International Journal of System Assurance Engineering and Management* 32: 1–12.

12 Tang, Y., Cheng, N., Wu, W. et al. (2019). Delay-minimization routing for heterogeneous VANETs with machine learning based mobility prediction. *IEEE Transactions on Vehicular Technology* 68 (4): 3967–3979.

13 Tong, J., Gu, X., Zhang, M. et al. (2021). Traffic flow prediction based on improved SVR for VANET. *2021 Fourth International Conference on Advanced Electronic Materials, Computers and Software*

Engineering (AEMCSE), pp. 402–405, Dublin, Ireland (4–5 March 2021). UK: IEEE.

14 Hua, Y., Guo, J., and Zhao, H. (2015). Deep belief networks and deep learning. *Proceedings of 2015 International Conference on Intelligent Computing and Internet of Things*, pp. 1–4, Harbin (25 May 2015). UK: IEEE.

15 Kuremoto, T., Kimura, S., Kobayashi, K., and Obayashi, M. (2014). Time series forecasting using a deep belief network with restricted Boltzmann machines. *Neurocomputing* 137: 47–56.

16 Hashim, F.A., Houssein, E.H., Mabrouk, M.S. et al. (2019). Henry gas solubility optimization: a novel physics-based algorithm. *Future Generation Computer Systems* 101: 646–667.

17 Yıldız, B.S., Yıldız, A.R., Pholdee, N. et al. (2020). The Henry gas solubility optimization algorithm for optimum structural design of automobile brake components. *Materials Testing* 62 (3): 261–264.

18 Zhang, T., Ding, W., Chen, T. et al. (2021). A graph convolutional method for traffic flow prediction in highway network. *Wireless Communications and Mobile Computing* 2021: 1–8.

4

Crow Search Algorithm Based Vector Quantization Approach for Image Compression in 6G Enabled Industrial Internet of Things Environment

Maha M. Althobaiti

Department of Computer Science, College of Computing and Information Technology, Taif University P.O.Box 11099, Taif 21944, Saudi Arabia

4.1 Introduction

The adaption of emergent applications and technological trends of the Internet of Things (IoT) in the industrial system is dominant toward the expansion of Industrial IoT (IIoT) [1]. IIoT assists as a novel form of IoT in the industrial field by systematizing smart objects for communicating, sensing, collecting, and processing real-time events in industrial systems. Generally, IIoT is determined by IoT as it is applied between distinct types of industry including logistics, transportation, energy or utilities, manufacturing (Industry 4.0), oil, mining, gas, aviation, metals, etc. [2, 3]. The advancement in smart environment, home automation, sensor networks, and smart cities for military and civilian applications has been incorporated for designing IoT. With current development in the field of remote sensor techniques, it is easier for capturing high quality remote sensing images with distinct satellites along with sensing devices and is certainly beneficial to remote sensing image application. But the massive number of data, i.e. remote sensing images, became a complex problem for transmission and storage [4]. Consequently, compression technique became indispensable in processing the remote

AI-Enabled 6G Networks and Applications, First Edition. Edited by Deepak Gupta, Mahmoud Ragab, Romany Fouad Mansour, Aditya Khamparia, and Ashish Khanna.

sensing image. In general, traditional compression method is applied for compressing remote sensing images, JPEG2000, and improved standard. However, the remote sensing image holds distinct features including lucid data, precise geographical matter, complex spatial data, etc. It is implicit that this traditional model fails to treat high frequency data appropriately and only focuses on maintaining several low frequency information that could not be beneficial [5]. In order to compress remote sensing images, this feature must be taken into account at the time of the compression method. Figure 4.1 depicts the uses of IIoT technique.

Vector quantization (VQ) approach outperformed other techniques including differential PCM (DPCM), pulse code modulation (PCM), and adoptive DPCM that belongs to the class of scalar quantization

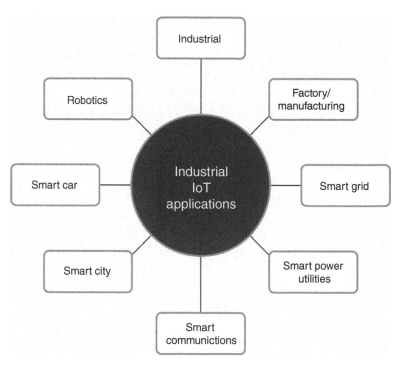

Figure 4.1 Applications of IIoT.

method. VQ [6] is the more commonly known lossy image compression method that is mainly a c-means clustered method broadly employed for pattern recognition and speech recognition [7], image compression, image coding, and face detection speech due to its benefits that include simple decoding framework and higher compression rate that offers lower distortion. Research involves the relation of evolutionary optimization algorithm with the Linde–Gray–Buzo (LGB) approach [8–10]. In Ref. [11], a new compression method is proposed. First, it spectrally decorrelated the image with principal component analysis (PCA) and VQ, later employing JPEG2000 to the principal component (PC) exploiting spatial correlation for compression. Next, exploit the fact that reduction dimension conserves large data in the initial component and allocates more depth to the initial PC. Lu et al. [12] presented a deep convolutional neural network (DCNN) framework for image compression, whereby the decoder, encoder, and quantizer are collectively learned. Especially, a fully convolution vector quantization network (VQNet) was presented for quantizing the feature vector of the image illustration, whereby the representor vector of VQNet is collectively enhanced by other network variables via end-to-end training. Even though many present DCNN-based methods have been trained on large-scale dataset, further we implement fine-tuning of the encoder and the code created by the VQNet on the input image for additionally improving efficiency of the compression. Figure 4.2 showcases the components in service aware 6G networks.

Zhou et al. [13] presented two effective compression methods for digital images with an adoptive selection method for image inpainting, VQ, and side match vector quantization (SMVQ). On the transmitter side, afterward, the novel image is separated into blocks, the compression can be performed block by block. In this system, block in prestated location is compressed initially by VQ. For all the residual blocks, the ideal compression technique (for the initial system, together with inpainting or VQ, and the next system, embraces inpainting, VQ, and SMVQ) can be described by calculating the mean square error (MSE) among the original block and in painted outcome and compared to a predetermined threshold.

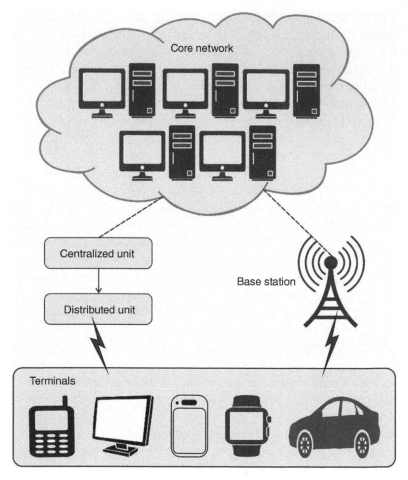

Figure 4.2 Components in service aware 6G networks.

In Ref. [14], a VQ approach to encode the bi-orthogonal wavelet decomposed images with hybrid pattern search optimization algorithm (hADEPS) and adoptive differential evolution (ADE) is presented. ADE is an adopted form of differential evolution (DE) where mutation process is created adoptive according to the fitness value or ascending or descending objective functions and verified on 12 arithmetical standard functions. In Ref. [15], an approach named integrated development environment (IDE)-LBG is presented that employs enhanced DE approach in addition to Linde–Buzo–Gray (LBG) for making optimal VQ codebook. The presented IDE effectively operates when compared to the conventional DE with modification in the boundary control mechanism and the scaling factor. The IDE makes good solution by effective exploitation and exploration of the searching space. Next, the optimum solution attained by the IDE is presented as the primary codebook for the LBG.

This study introduces a novel crow search algorithm based vector quantization approach for image compression in 6G enabled IIoT environment, called CSAVQ-ICIIoT model. The proposed CSAVQ-ICIIoT model intends to accomplish effectual image compression by optimizing codebook construction process in 6G enabled IIoT platform. The CSAVQ-ICIIoT technique includes LBG with VQ technique for image compression. Besides, the optimal codebook construction process is performed by the use of crow search algorithm (CSA). For examining the improved performance of the CSAVQ-ICIIoT model, a detailed result analysis is made and the results are inspected under several measures.

4.2 The Proposed Model

This study has developed a new CSAVQ-ICIIoT model for accomplishing effectual image compression by optimizing codebook construction process in 6G enabled IIoT platform. The CSAVQ-ICIIoT technique includes LBG with VQ technique for image compression. Besides, the optimal codebook construction process is performed by the use of CSA and thereby improves the compression efficiency.

4.2.1 Overview of VQ

VQ is a lossy data compression method in block coding. The codebook generation is called the essential procedure of VQ. Assume the size of novel image $Y = \{y_{ij}\}$ be $M \times M$ pixel that divides different blocks with size of $n \times n$ pixels [16]. In another word, there is $N_b = \left\lceil \dfrac{N}{n} \right\rceil \times \left\lceil \dfrac{N}{n} \right\rceil$ block that represents group of input vectors $= (x_i, i = 1, 2, ..., N_b)$. Consider L as $n \times n$. The input vector χ_i, $\chi_i \in \mathfrak{R}^L$ whereas $\mathfrak{R}^L$ is L-dimension Euclidean space. A codebook C encompasses N_c L-dimension codeword, that is $= \{c_1, c_2, ..., c_{N_c}\}$, $c_j \in \mathfrak{R}^L$, $\forall j = 1, 2, ..., N_c$. All the input vectors are denoted as a row vector $\chi_i = (x_{i1}, x_{i2}, ..., x_{iL})$ and the jth codeword of codebook is represented by $c_j = (c_{j1}, c_{j2}\ c_{jL})$. The VQ technique assigns input vector to an associated codeword, and that would replace the related input vector lastly to attain the objective of compression. The optimization of C in terms of MSE is expressed by minimalizing the distortion function D. Generally, lesser the value of D is, the greater the quality of C would be.

$$D(C) = \frac{1}{N_b} \sum_{j=1}^{N_c} \sum_{i=1}^{N_b} \mu_{ij} \cdot \left\| x_i - c_j \right\|^2 \tag{4.1}$$

subjected to the subsequent limitations:

$$\sum_{j=1}^{N_c} \mu_{ij} = 1, \quad \forall i \in \{1, 2, ..., N_b\} \tag{4.2}$$

$$\mu_{ij} = \begin{cases} 1 \ if \ \chi_i \ is \ in \ the \ j\text{th cluster}; \\ 0 \ \text{otherwise} \end{cases} \tag{4.3}$$

also

$$L_k \leq c_{jk} \leq U_k, \quad k = 1, 2, ..., L \tag{4.4}$$

Where L_k represents the minimal of kth element in each training vector, and U_k denotes the maximal of kth element in each input vector. $\|x - c\|$ represents the Euclidean distance among the vector χ and codeword c.

Two essential conditions exist for optimum vector quantizer.

1) The partition R_j, $j = 1, ..., N_c$ should fulfill

$$R_j \supset \{x \in \chi : d(x, c_j) < d(x, c_k), \quad \forall k \neq j\}. \tag{4.5}$$

2) The codeword c_j should be provided by the centroid of R_j:

$$c_j = \frac{1}{N_j} \sum_{i=1}^{N_j} x_i, \quad x_i \in R_j, \tag{4.6}$$

Where N_j indicates the overall amount of vector belongs to R_j.

4.2.2 LBG Model

This approach is called as generalized Lloyd algorithm (GLA) or LBG. It employs the two aforementioned conditions to input vector for defining the codebook.

Assume input vector, x_i, $i = 1, 2, ..., N_b$, distance function d, and primary codeword $c_j(0)$, $j = 1, ..., N_c$. The LBG employs the two conditions to generate the optimum codebook:

1) Partition the input vector into various groups with the minimal distance rule. The resultant partition is saved in a $N_b \times N_c$ binary indicator matrix U element is determined by:

$$\mu_{ij} = \begin{cases} 1 \ if \ d(x_i, c_j(k)) = \min \ d(x_i, c_p) \\ 0 \ otherwise \end{cases} \tag{4.7}$$

2) Define the centroid of all the partitions. Replace the old codeword with this centroid:

$$c_j(k+1) = \frac{\sum_{i=1}^{N_b} \mu_{ij} x_i}{\sum_{i=1}^{N_b} \mu_{ij}}, \quad j = 1, ..., N_c. \tag{4.8}$$

3) Repeat steps (1) and (2) till no $c_j, j = 1, ..., N_c$ variations to any further extent.

4.2.3 Process Involved in CSAVQ-ICIIoT Model

Crows are regarded as the smartest creatures among birds. They have big brains compared to body size. As per the brain-to-body relationship, its brain is slightly less when compared to humans. They have the capacity to make tools and showed self-knowledge in mirror tests. They are aware of each other once the threat occurred and memorize faces [17]. Furthermore, they share data, apply tools in complicated manner, and remember their secret food position. They observe various birds, trace where the bird kept the food in secret, and steal the food once the bird has left its position. After performing robbery, they play it safe, e.g. changing hiding spot for avoiding a victim after this. They utilize their knowledge to determine the remaining thief and secured manner of defending their food from getting stolen by others [17].

The following properties show the standard of crows:

- Crows keep in mind the position of a secret location.
- Crows live as a group.
- Crows secure their hideout from the possibility of being stolen.
- Crows follow each other to perform theft.

Apparently, there are N-dimensional environments encompassing various crows. The overall crows represent C and the position of crow u in time (iteration) *iter* in the searching space (SS) is defined as follows:

$$V^{u,iter}(p = 1, 2, ..., C; iter = 1, 2, ...,) \, iter_{max} \tag{4.9}$$

Where $V^{u,iter} = \left[V_1^{u,iter}, V_2^{u,iter}, ..., V_c^{u,iter} \right]$ and $iter_{max}$ denotes the iteration with maximal amount. All the crows have a memory in which the position of a secret location was stored. At the time of iteration, the position of secret place of crow u is characterized as $s^{u,\ iter}$. That is the best position crow u accomplished until now. In the memory of all the crows, the position of improved experience was saved. Crow starts searching for good sources in the environment.

Assume that in iteration, crow v must go to its secret position, $s^{v,\ iter}$. In iteration, crow u determined for tracing crow v to the secret position of crow v. Two events might take place in this phase,

Event 1: Crow v has no idea that crow u is tracing it. Subsequently, crow u reaches the secret position of crow v.

$$V^{u,iter+1} = V^{u,iter} + k_j \times fll^{u,iter} \times \left(S^{v,iter} - V^{u,iter} \right) \tag{4.10}$$

Where k_j denotes an arbitrary value with standard distribution among 0 and 1, and $fll^{u,iter}$ represents the flight length of crow u in iteration.

Minimal value of *fll* leads to local search and maximal value results in global search.

Event 2: Crow v knows that crow u was tracing it. Therefore, to protect its secret position from theft, crow v deceives crow u by moving to another position of the *SS*.

$$V^{u,iter+1} = \begin{cases} V^{u,iter} + k_j \times fll^{u,iter} \times \left(S^{v,iter} - V^{u,iter}\right) k_j \geq AWP^{v,iter} \\ a\ random\ location\ otherwise \end{cases} \quad (4.11)$$

In which $AWP^{v,iter}$ represents the possibility of awareness of crow v in iteration.

The total procedure of CSAVQ-ICIIoT technique was brief as to five subprocesses that are listed here. Initially, the parameter is endured initialized in which the generated codebook by LBG technique has selected to primary solutions [18]. After the initialized method, the present optimum solutions are selected as resolved of the fitness of all the place and defines maximal fitness place as finest one. If the current optimum solution is derived, novel solutions are created under the roaming process. Afterward, a ranking technique has introduced and the solution is ranked dependent upon the feed forward (FF) under the successive step and selects the optimum one.

4.3 Results and Discussion

The proposed model is simulated using the benchmark dataset [19]. The dataset holds high resolution remote sensing images. In this work, a set of six images with the dimension of 1024×1024 and $3072\ kB$ size is taken. A few sample images are illustrated in Figure 4.3.

A brief compression ratio (CR) examination of the CSAVQ-ICIIoT model with existing models under distinct images is portrayed in Table 4.1 and Figure 4.4. The results indicated that the CSAVQ-ICIIoT model has resulted in lower values of CR over the other methods. For instance, on image 1, the CSAVQ-ICIIoT model has obtained minimal CR of 0.0735 whereas the DCNN, BTOT, and JPEG techniques have attained maximum CR of 0.0872, 0.1126, and 0.1250, respectively. In

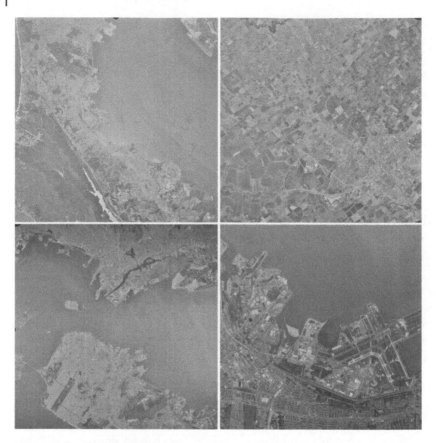

Figure 4.3 Sample images.

addition, on image 3, the CSAVQ-ICIIoT technique has obtained lesser CR of 0.0846 whereas the DCNN, BTOT, and JPEG techniques have attained maximum CR of 0.0957, 0.1283, and 0.1338, respectively. Moreover, on image 6, the CSAVQ-ICIIoT approach has obtained lower CR of 0.0946 whereas the DCNN, BTOT, and JPEG techniques have attained higher CR of 0.1071, 0.1341, and 0.1543 correspondingly.

Table 4.2 and Figure 4.5 illustrate the compression factor (CF) examination of the CSAVQ-ICIIoT model with existing techniques.

Table 4.1 Compression ratio analysis of CSAVQ-ICIIoT technique with different images.

Test images	Compression ratio (CR)			
	CSAVQ-ICIIoT	DCNN	BTOT	JPEG
Image 1	0.0735	0.0872	0.1126	0.1250
Image 2	0.1048	0.1113	0.1527	0.1611
Image 3	0.0846	0.0957	0.1283	0.1338
Image 4	0.0937	0.1064	0.1523	0.1533
Image 5	0.1184	0.1296	0.1934	0.1969
Image 6	0.0946	0.1071	0.1341	0.1543

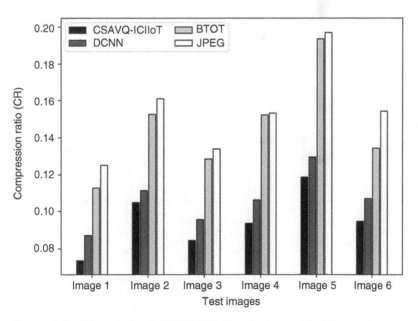

Figure 4.4 CR analysis of CSAVQ-ICIIoT technique with different images.

Table 4.2 Compression factor analysis of CSAVQ-ICIIoT technique with different images.

Test images	Compression factor (CF)			
	CSAVQ-ICIIoT	DCNN	BTOT	JPEG
Image 1	12.9505	11.4627	8.8786	8.0000
Image 2	10.3662	8.982 50	6.5501	6.2061
Image 3	11.9377	10.4490	7.7970	7.4745
Image 4	10.3789	9.394 50	6.5641	6.5223
Image 5	09.3766	7.718 60	5.1717	5.0777
Image 6	10.3764	9.337 40	7.4563	6.4810

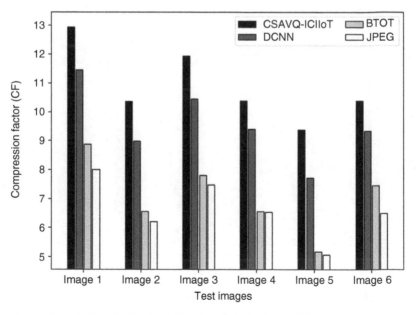

Figure 4.5 CF analysis of CSAVQ-ICIIoT technique with different images.

The results demonstrated that the CSAVQ-ICIIoT model has the capability of accomplishing maximum compression efficiency over the other techniques. For instance, under image 1, the CSAVQ-ICIIoT model has offered higher CF of 12.9505 whereas the DCNN, BTOT, and JPEG techniques have resulted to lower CF of 11.4627, 8.8786, and 8.0000. Simultaneously, under image 3, the CSAVQ-ICIIoT model has obtainable maximal CF of 11.9377 whereas the DCNN, BTOT, and JPEG systems have resulted in minimal CF of 10.4490, 7.7970, and 7.4745. Concurrently, under image 6, the CSAVQ-ICIIoT technique has offered higher CF of 10.3764 whereas the DCNN, BTOT, and JPEG methods have resulted in minimum CF of 9.337 40, 7.4563, and 6.4810.

A detailed basic resource (BR) examination of the CSAVQ-ICIIoT technique with existing approaches under various images is depicted in Table 4.3 and Figure 4.6. The results indicated that the CSAVQ-ICIIoT technique has resulted to lower values of BR over the other methods. For instance, on image 1, the CSAVQ-ICIIoT model has obtained minimal BR of 1.0387 whereas the DCNN, BTOT, and JPEG approaches have gained increased BR of 2.0938, 2.7031, and 3.0000 correspondingly.

Followed by, on image 3, the CSAVQ-ICIIoT system has achieved minimal BR of 2.1038 whereas the DCNN, BTOT, and JPEG techniques

Table 4.3 Bit rate analysis of CSAVQ-ICIIoT technique with different images.

Test images	Bit rate			
	CSAVQ-ICIIoT	DCNN	BTOT	JPEG
Image 1	1.0387	2.0938	2.7031	3.0000
Image 2	2.4332	2.6719	3.6641	3.8672
Image 3	2.1038	2.2969	3.0781	3.2109
Image 4	2.0937	2.5547	3.6563	3.6797
Image 5	2.9846	3.1094	4.6406	4.7266
Image 6	2.3210	2.5703	3.2188	3.7031

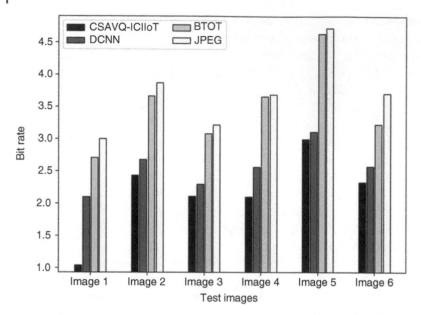

Figure 4.6 BR analysis of CSAVQ-ICIIoT technique with different images.

have attained superior BR of 2.2969, 3.0781, and 3.2109 correspondingly. Eventually, on image 6, the CSAVQ-ICIIoT methodology has gained decreased BR of 2.3210 whereas the DCNN, BTOT, and JPEG techniques have attained increased BR of 2.5703, 3.2188, and 3.7031 correspondingly.

A comparative MSE examination of the CSAVQ-ICIIoT model with existing models under distinct images is portrayed in Table 4.4 and Figure 4.7. The results indicated that the CSAVQ-ICIIoT model has resulted to lower values of MSE over the other methods. For instance, on image 1, the CSAVQ-ICIIoT model has obtained reduced MSE of 0.75 whereas the DCNN, BTOT, and JPEG approaches have attained superior MSE of 1.40, 2.80, and 6.80, respectively. Besides image 3, the CSAVQ-ICIIoT technique has attained minimal MSE of 0.89 whereas the DCNN, BTOT, and JPEG techniques have attained enhanced MSE of 1.60, 2.70, and 24.90, respectively. At last, on image 6, the

Table 4.4 MSE analysis of CSAVQ-ICIIoT technique with different images.

Test images	Mean square error (MSE)			
	CSAVQ-ICIIoT	DCNN	BTOT	JPEG
Image 1	0.75	1.40	2.80	06.80
Image 2	0.88	2.10	3.90	12.30
Image 3	0.89	1.60	2.70	24.90
Image 4	0.74	0.80	1.20	19.70
Image 5	0.77	2.90	3.20	22.30
Image 6	0.83	2.70	3.80	31.90

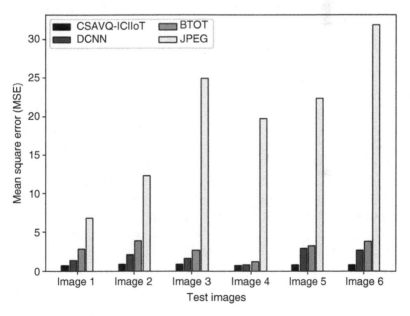

Figure 4.7 MSE analysis of CSAVQ-ICIIoT technique with different images.

Table 4.5 PSNR analysis of CSAVQ-ICIIoT technique with different images.

	Peak signal noise ratio (PSNR)			
Test images	CSAVQ-ICIIoT	DCNN	BTOT	JPEG
Image 1	49.38	47.40	45.90	43.97
Image 2	48.69	46.52	45.18	42.68
Image 3	48.64	47.11	45.97	41.15
Image 4	49.44	48.62	47.73	41.66
Image 5	49.27	45.82	45.61	41.39
Image 6	48.94	45.97	45.23	40.61

CSAVQ-ICIIoT model has obtained lesser MSE of 0.83 whereas the DCNN, BTOT, and JPEG techniques have attained increased MSE of 2.70, 3.80, and 31.90 correspondingly.

Table 4.5 and Figure 4.8 demonstrate the PSNR examination of the CSAVQ-ICIIoT approach with existing techniques [20]. The outcomes outperformed that the CSAVQ-ICIIoT method has the capability of accomplishing maximum compression efficiency over the other techniques. For instance, under image 1, the CSAVQ-ICIIoT technique has obtainable higher PSNR of 49.38 whereas the DCNN, BTOT, and JPEG algorithms have resulted to lower PSNR of 47.40, 45.90, and 43.97.

Concurrently, under image 3, the CSAVQ-ICIIoT model has offered higher PSNR of 48.64 whereas the DCNN, BTOT, and JPEG techniques have resulted in lesser PSNR of 47.11, 45.97, and 41.15. Finally, under image 6, the CSAVQ-ICIIoT technique has offered maximum PSNR of 48.94 whereas the DCNN, BTOT, and JPEG approaches have resulted to lower PSNR of 45.97, 45.23, and 40.61.

4.4 Conclusion

This study has developed a new CSAVQ-ICIIoT model for accomplishing effectual image compression by optimizing codebook construction process in 6G enabled IIoT platform. The CSAVQ-ICIIoT technique

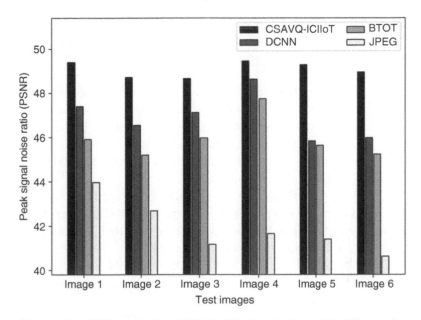

Figure 4.8 PSNR analysis of CSAVQ-ICIIoT technique with different images.

includes LBG with VQ technique for image compression. Besides, the optimal codebook construction process is performed by the use of CSA and thereby improves the compression efficiency. For examining the improved performance of the CSAVQ-ICIIoT model, a detailed result analysis is made and the results are inspected under several measures. The experimental results reported the enhanced outcomes of the CSAVQ-ICIIoT model over the other methods. In future, it can be deployed in real-time unmanned aerial vehicle networks.

References

1 Nguyen, D.C., Ding, M., Pathirana, P.N. et al. (2021). 6G internet of things: a comprehensive survey. *IEEE Internet of Things Journal* 9 (1): 359–383.

2 Gong, Y., Yao, H., Wang, J. et al. (2022). Edge intelligence-driven joint offloading and resource allocation for future 6G industrial internet of things. *IEEE Transactions on Network Science and Engineering* 21: 1–12.

3 Gao, Y. (2021). Using artificial intelligence approach to design the product creative on 6G industrial internet of things. *International Journal of System Assurance Engineering and Management* 12: 696–704.

4 Chen, S., Wang, Z., Zhang, H. et al. (2019). Fog-based optimized kronecker-supported compression design for industrial IoT. *IEEE Transactions on Sustainable Computing* 5 (1): 95–106.

5 Park, J., Park, H., and Choi, Y.J. (2018). Data compression and prediction using machine learning for industrial IoT. *2018 International Conference on Information Networking (ICOIN)*, pp. 818–820, Chiang Mai, Thailand (10–12 Jan 2018). Thailand: IEEE.

6 Ammah, P.N.T. and Owusu, E. (2019). Robust medical image compression based on wavelet transform and vector quantization. *Informatics in Medicine Unlocked* 15: 100183.

7 Hilles, S.M. (2018). Sofm and vector quantization for image compression by component. *2018 International Conference on Smart Computing and Electronic Enterprise (ICSCEE)*, pp. 1–6, Shah Alam, Malaysia (11–12 July 2018). Malaysia: IEEE.

8 Elsayad, A.M. (2021). Medical Image Compression Using Vector Quantization and Gaussian Mixture Model.(Dept. E). *MEJ. Mansoura Engineering Journal* 28 (3): 13–21.

9 Khan, M.M. (2021). An implementation of vector quantization using the genetic algorithm approach. *arXiv* preprint arXiv:2102.08893.

10 Suguna, T. and Shanmugalakshmi, R. (2021). Secure image communication through adaptive deer hunting optimization based vector quantization coding of perceptually encrypted images. *Wireless Personal Communications* 116 (3): 2239–2260.

11 Báscones, D., González, C., and Mozos, D. (2018). Hyperspectral image compression using vector quantization, PCA and JPEG2000. *Remote Sensing* 10 (6): 907.

12 Lu, X., Wang, H., Dong, W. et al. (2019). Learning a deep vector quantization network for image compression. *IEEE Access* 7: 118815–118825.

13 Zhou, Q., Yao, H., Cao, F., and Hu, Y.C. (2019). Efficient image compression based on side match vector quantization and digital inpainting. *Journal of Real-Time Image Processing* 16 (3): 799–810.

14 Chiranjeevi, K. and Jena, U. (2018). SAR image compression using adaptive differential evolution and pattern search based K-means vector quantization. *Image Analysis and Stereology* 37 (1): 35–54.

15 Nag, S. (2019). Vector quantization using the improved differential evolution algorithm for image compression. *Genetic Programming and Evolvable Machines* 20 (2): 187–212.

16 Geetha, K., Anitha, V., Elhoseny, M. et al. (2021). An evolutionary lion optimization algorithm-based image compression technique for biomedical applications. *Expert Systems* 38 (1): e12508.

17 Meraihi, Y., Gabis, A.B., Ramdane-Cherif, A., and Acheli, D. (2021). A comprehensive survey of Crow Search Algorithm and its applications. *Artificial Intelligence Review* 54 (4): 2669–2716.

18 Rani, M.L.P., Rao, G.S., and Rao, B.P. (2021). An efficient codebook generation using firefly algorithm for optimum medical image compression. *Journal of Ambient Intelligence and Humanized Computing* 12 (3): 4067–4079.

19 https://sipi.usc.edu/database/.

20 Sujitha, B., Parvathy, V.S., Lydia, E.L. et al. (2021). Optimal deep learning based image compression technique for data transmission on industrial Internet of things applications. *Transactions on Emerging Telecommunications Technologies* 32 (7): e3976.

5

Design of Artificial Intelligence Enabled Dingo Optimizer for Energy Management in 6G Communication Networks

Pooja Singh[1,2], Marcello Carvalho dos Reis[3,4], and Victor Hugo C. de Albuquerque[5]

[1] Graduate Program in Telecommunication Engineering, Federal Institute of Education, Science and Technology of Ceará, Fortaleza, Brazil
[2] Department of Computer Science and Engineering, GL Bajaj Institute of Technology and Management, Greater Noida, Uttar Pradesh, India
[3] Graduate Program in Telecommunication Engineering, Federal Institute of Education, Science and Technology of Ceará, Fortaleza, Brazil
[4] Meteora, Fortaleza, Brazil
[5] Department of Teleinformatics Engineering, Federal University of Ceará, Fortaleza, Brazil

5.1 Introduction

The sixth generation (6G) network should give preferable execution over past generations to meet the necessities of emerging administrations and applications. The evolution of 6G network is shown in Figure 5.1. To lighten the growing energy constraints toward 6G, the scholarly world and industry have led broad exploration. Furthermore, the accessible answers to address the huge energy utilization primarily comes from two sections: energy-proficient network design [1] and energy harvesting [2]. In particular, energy harvesting units, such as the sun powered chargers, wind turbines, and vibration reaper, are generally embraced to change the different sorts of energy over to power for the specialized gadgets [3]. Energy proficiency is one more interesting point for future remote networks. It is attractive to have equipment that is viable with the energy prerequisites of 6G. The energy change started

AI-Enabled 6G Networks and Applications, First Edition. Edited by Deepak Gupta, Mahmoud Ragab, Romany Fouad Mansour, Aditya Khamparia, and Ashish Khanna.
© 2023 John Wiley & Sons Ltd. Published 2023 by John Wiley & Sons Ltd.

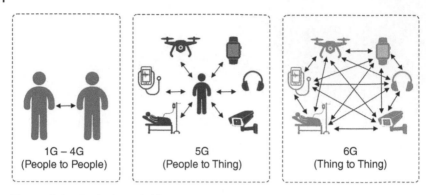

Figure 5.1 Evolution of 6G networks.

by the political arrangement and the increasing effect of sustainable power utilization in land require an extra viable idea of nearby energy creation and energy utilization.

Balancing energy utilization between energy creation and acquisition of energy must be accomplished by a productive, solid, and practical energy management framework. Heat energy for heating and high temperature water and electrical energy should be utilized to control the buildings through ideas with decreased CO_2 emanations [4]. Likewise, numerous different assignments in an advanced property should have the option to settle expense successfully with the assistance of digital parts [5].

Neighborhood date generation can be managed by software stages with quick connects to various Internet of Things (IoT) radio technologies. Different domains involved in 6G networks are depicted in Figure 5.2. Energy management systems in light of enterprise IT programming design permit customers and market players to create new administrations like energy management [5] for little units (offices, lofts, creation offices), energy management for entire buildings and offices, energy management for metropolitan areas, and industry parks with information estimation and metering. Digital control for client items and applications manages own prerequisites connected with

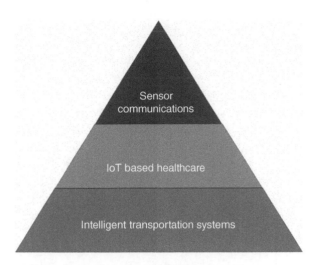

Figure 5.2 Domains in 6G networks.

energy administrations and items consequently likewise using market interfaces and simultaneously support the quality and security of supply of the grid-based power framework [6].

With the AIDO, gadgets have an elective wellspring of force from remote correspondence. This is therefore extending the battery duration of gadgets. The purpose of 6G driven energy effective network in box (NIB) is to reshape the actual world into brilliant and inescapable. Also, NIB is the vital enabling substance to foster self-organizing and consistent association with better and reasonable administrations for example, interactive media (i.e. text, sound, and video, etc.) [7]. The 6G enabled NIB furnishes adaptable and versatile foundation with less upward and helpful substance conveyance for far-off modern robotization. The handheld remote stage gives emerging open doors to modern computerization, disseminated frameworks, and vehicles with networked installed framework. Due to hard real-time nature with severe cutoff times, modern applications needs delay open-minded and blunder-free correspondence. Nowadays different applications, for example, cell, earthbound, and marine, are very much covered by NIB technologies [8, 9].

Hu et al. [10] presented an energy-proficient in-network registering worldview for 6G that incorporates network capacities into an overall processing stage as opposed to assigning figuring undertakings to network gadgets. The figuring stage that coordinates network capacities replaces conventional network gadgets and goes about as a network hub. Not at all like customary network gadgets, the network hub gives a brought together working climate to application undertakings through hypervisors and compartments. Sodhro et al. [11] propose an original machine learning-driven portability, the board strategy for the proficient correspondence in modern NIB applications. Second, an original engineering of 6G-based canny quality of energy (QoE) and quality of service (QoS) enhancement in modern NIB is proposed.

In Ref. [7], a fuzzy rationale-based wakeup procedure is proposed, which completely considers the energy wakeup level and the accessible network asset proportion. Then, at that point, the sun based energy states are dissected numerically by utilizing the dispersion guess technique. At long last, to keep basic signals (BSs) from exchanging much of the time between the dozing and nonsleeping model, the enlivening edge is upgraded by the punishment work strategy. Mao et al. [12] proposed artificial intelligence (AI) based versatile security particular strategy for 6G IoT networks where the IoT gadgets are associated with cell networks by means of various recurrence groups, including terahertz (THz) and millimeter wave (mmWave). The IoT detecting gadgets are accepted to help the energy gathering procedure, which is relied upon to be broadly taken on in 6G. In our proposition, the extended Kalman filter (EKF) technique is first taken on to foresee future collecting power. In Ref. [13], a new multidimensional-intelligent multiple access (MD-IMA) method is proposed in this paper to take advantage of unique asset imperatives among heterogeneous hardware for 5G past and 6G networks. With the help of continuous information examination, ongoing QoS necessities and asset accessibility of the connected gear are not set in stone in the proposed MD-IMA.

This paper presents an AI enabled dingo optimizer for energy management (AIDO-EM) in 6G networks. The presented AIDO-EM technique involves the major goal of minimizing the energy utilization and maximizing the lifetime of the 6G enabled IoT devices. For accomplishing this, a new dingo optimization algorithm (DOA) is applied for cluster enabled routing to achieve effective data distribution among the devices and choose effective gateway heads (GWH). In order to demonstrate the enhanced outcomes of the AIDO-EM technique, a series of simulations were carried out and the results reported the enhanced performance over its recent state-of-art approaches.

5.2 The Proposed Model

In this study, a novel AIDO-EM technique has been developed to achieve effective energy efficiency in the 6G networks. The presented AIDO-EM technique intends to accomplish minimal energy usage and maximum lifetime of the 6G enabled IoT devices. In addition, a novel DOA is applied for cluster enabled routing to achieve effective data distribution among the devices and choose effective GWHs.

5.2.1 Process Involved in DOA

The DOA is a bio-inspired technique for global optimization that stimulates hunting strategy of dingoes. This strategy is attacking through scavenging, persecution, and grouping tactics behavior [14, 15]. The Australian dingo dog is at risk of extinction. Hence, survival possibility of dingoes is taken into account.

Level 1: Group attack. Usually, dingoes gather in group while hunting. They can detect the position of the prey and surround it. This can be expressed as:

$$\vec{x}_i(t+1) = \beta_1 \sum\nolimits_{k=1}^{na} \frac{\left[\overrightarrow{\varphi_k(t)} - \vec{x}_i(t)\right]\frac{1}{\phi_k(t)}}{na} - \vec{x}_{i*}(t) \tag{5.1}$$

Where $\vec{x}(t+1)$ denotes the location of searching agent (shows dingoes' motion), na indicates an arbitrary integer value within $[2, \frac{SizePop}{2}]$, whereas SizePop denotes the overall size of the population. $\overrightarrow{\varphi_k(t)}$ is a subset of searching agents (dingoes that would attack) here $\phi \subset X$, X

denotes the population arbitrarily created, $\vec{x}_i(t)$ indicates the present searching agent, $\vec{x}_i(t)$ shows the optimal searching agent found from the preceding iteration, β_1 indicates an arbitrary value created within $[-2, 2]$; it is a scaling factor that alters the magnitude and sense of dingoes' trajectory.

Level 2: Persecution. Usually, dingoes hunt smaller prey and rush till the prey is individually fixed as follows:

$$\vec{x}_i(t+1) = \vec{x}_*(t) + \beta_1 * e^{\beta_2} * \left(\vec{x}_{r_1}(t) - \vec{x}_i(t)\right) \tag{5.2}$$

Where $\vec{x}(t+1)$ shows the dingoes' motion, $\vec{x}_i(t)$ indicates the present searching agent, $\vec{x}_*(t)$ denotes the optima; searching agent initiates from the preceding round, β_1 involves identical value, β_2 signifies an arbitrary value within $[-1, 1]$, r_1 denotes the arbitrary value within 1 to the size of maximal of searching agent (dingo), and $\vec{x}_{r1}(t)$ indicates the r_1-th searching agent selected, where $i \neq r_1$.

Level 3: Scavenger. It is determined as the action once dingoes detect carrion to eat, once they are walking arbitrarily in the habitat.

$$\vec{x}_i(t+1) = \frac{1}{2}\left[e^{\beta_2} * \vec{x}_{r_1}(t) - \left(-1\right)^{\sigma} * \vec{x}_i(t)\right] \tag{5.3}$$

where $\vec{x}(t+1)$ indicates the dingoes' motion, β_2 has the same value as in Eq. (5.2), r_1 indicates the arbitrary value within 1 to the size of maximal of searching agent, $\vec{x}_{r_1}(t)$ represents the r_1-th selected searching agent, $\vec{x}_i(t)$ denotes the present searching agent, whereby $i \neq r_1$, and σ indicates a binary value generated arbitrarily.

Level 4: Dingo survival rates. The dingo existence rate value is shown as follows:

$$Survival\,(i) = \frac{fitness_{\max} - fitness(i)}{fitness_{\max} - fitness_{\min}} \tag{5.4}$$

In which, $fitness_{\max}$ and $fitness_{\min}$ denote the worse and optimal fitness values in the existing generations, correspondingly, whereby *fitness*(*i*) represents the present fitness value of *i*-th searching agent. The existence vector comprises the normalized fitness within [0, 1].

$$\vec{x}_i(t) = \vec{x}_*(t) + \frac{1}{2}\left[\vec{x}_{r_1}(t) - \left(-1\right)^\sigma * \vec{x}_{r_2}(t)\right] \tag{5.5}$$

Where $\vec{x}_i(t)$ denotes the searching agent with lower survival rate that would be upgraded, r_1 and r_2 indicate arbitrary value within 1 to the maximal size of search, $\vec{x}_*(t)$ denotes the optimal searching agent found from the preceding iteration, and σ denotes binary value (dingoes), with $r_1 \neq r_2$, $\vec{x}_{r_1}(t)$ and $\vec{x}_{r_2}(t)$ denotes the r_1, r_2-th searching agent selected, and $\vec{x}_*(t)$ denotes the optimal searching agent found from the preceding iteration.

5.2.2 Steps Involved in Energy Management Scheme

The main objective is to decrease the power utilization of 6G-assisted IoT network and thus, developing green transmission among the network. The GWH performs a significant role in managing the power utilization, especially in reduction of the number of communications in a network. The procedure of selecting GWH passes over the enhanced solution reduced by high wavelength spectrum with higher ordering (HWSHO). Then calculate the fitness function combined to distinct variables for selecting GWH. The fitness parameter describes individual fitness variable and consequence in the presented study [16]. GWH consume high energy when compared to the cluster member node, since they transfer information to the sink through gathering from them. Therefore, it becomes critical to choose that node as GWH that is augmented by the maximal probable value.

$$F_1 = \sum_{p=1}^{T_N} \frac{E_{rsd}(p)}{E_{in}(p)} \tag{5.6}$$

$$E_{rsd}(p) = E_{in}(p) - E_{txn} - E_{rxn} - E_{agg} \tag{5.7}$$

Because of the higher-energy condition for selecting node as *GH*, the initial fitness variable should be increased for the node selection. Once sensors are interactive with another node or sink, the transmission distance of node is of higher concern. In this fitness variable, assume the ratio of distance of certain candidate node and the average distance of node from the sink:

$$F_2 = \sum_{p=1}^{T_N} \frac{D_{avg}}{D_{NS}(p)} \tag{5.8}$$

$$D_{avg} = \frac{1}{T_N} \sum_{p=1}^{T_N} D_{NS}(p) \tag{5.9}$$

$$D_{NS} = \sqrt{(N_{x2} - N_{x1})^2 + (N_{y2} - N_{y1})^2} \tag{5.10}$$

Additionally, in Eq. (5.9), calculate the normal distance of node from the sink, and calculate the Euclidean distance among two nodes.

Once the network process begins, the energy of network reduces. Therefore, a moment comes once the amount of dead nodes get high, we represent the third fitness variable that states the condition of network remaining energy (NRE). For enhanced GWH election, the variable must be maximal:

$$F_3 = \frac{1}{T_N} \times \sum_{p=1}^{T_N} E_{rsd(p)} \tag{5.11}$$

When working with the 6G-assisted IoT device, there are different problems that delay the transmission among the sensors. The variable is the path-loss factor that should be taken into account for the data packet communication among the sensors. It assists in exploring the condition of the energy spending because of different wireless networks. The calculation of path loss is given as follows:

$$P_{TH}L(D_T) = P_L(T_{DST}) + 10 \times T_N \times \log\left(\frac{D_T}{T_{DST}}\right) \tag{5.12}$$

Moreover, $P_L(T_{DST})$ is calculated by:

$$P_L(T_{DST}) = 10 \times \log \times \left(4\pi \times Tim_{delay} \times Fq\right) \times c_L \tag{5.13}$$

Equation (5.14) is utilized for defining the inverse of calculated path loss through the distance D_T:

$$F_4 = \frac{1}{P_{TH}L(D_T)} \tag{5.14}$$

To accomplish the energy-effective GWH election, the fourth variable, F_4, must be increased. Once node of cluster consumes energy, the efficient value of cluster energy reduces. Therefore, we calculate the fifth fitness variable, that must be higher for selecting candidate node as GWH:

$$F_5 = \sum_{p=1}^{N_{cls}} E_{rsd(p)} \tag{5.15}$$

The multi-objective fitness function is transmuted to individual objective by incorporating different modules, that is fitness parameter, and multiply them to the linear weight function as follows:

$$F = \lambda \times F_1 + \sigma \times F_2 + \gamma \times F_3 + \beta \times F_4 + \rho \times F_5 \tag{5.16}$$

$$\lambda + \delta + \gamma + \sigma + \beta = 1 \tag{5.17}$$

The aforementioned equation provides the weighted summation of the distinct factors multiplied with parameter. Therefore, all the nodes are checked for fitness function. The node with maximum fitness value is chosen as generation hour (GH).

5.3 Experimental Validation

This section inspects the performance validation of the AIDO-EM model in terms of distinct dimensions with existing techniques [16]. Table 5.1 and Figure 5.3 provide a brief examination of the AIDO-EM model in terms of stability period, higher dimensions (HND), and lifetime. The results indicated that the AIDO-EM model has gained maximum performance over the other methods in terms of distinct measures. With respect to stability period, the AIDO-EM model has obtained higher stability period of 3362 rounds whereas the particle swarm

Table 5.1 Lifetime analysis of AIDO-EM model.

	No. of rounds		
Methods	Stability period	HND	Network lifetime
PSO-ECSM	930	1363	1656
R-LEACH	1363	2087	2552
CIRP	1587	2431	2949
ARSH-FATI	1966	2983	3621
Proposed model	3362	4983	6051

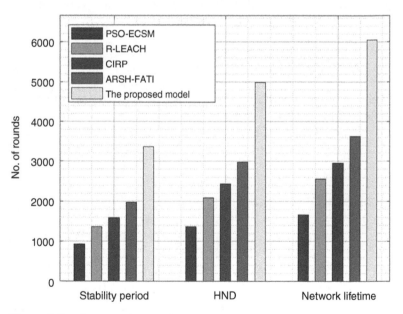

Figure 5.3 Comparative lifetime examination of AIDO-EM model.

optimization-based energy efficient clustering and sink mobility (PSO-ECSM), reliable low energy adaptive clustering hierachy (R-LEACH), centralized intelligent and resilient protection (CIRP), and ARSH-FATI methods have attained lower stability period of 930, 1363, 1587, and 1966 rounds, respectively. At the same time, with respect to HND, the AIDO-EM model has obtained higher stability period of 4983 rounds whereas the PSO-ECSM, R-LEACH, CIRP, and ARSH-FATI methods have attained lower stability period of 1363, 2087, 2431, and 2983 rounds, respectively. Moreover, with respect to lifetime, the AIDO-EM model has accomplished increased stability period of 6051 rounds whereas the PSO-ECSM, R-LEACH, CIRP, and ARSH-FATI methods have reached decreased stability period of 1656, 2552, 2949, and 3621 rounds, respectively.

The experimental result analysis of the AIDO-EM model with recent methods in terms of number of alive nodes (NOAN) is performed in Table 5.2 and Figure 5.4. The results indicated that the AIDO-EM model has reached maximum NOAN over the existing technique under all rounds. For instance, with 1000 rounds, the AIDO-EM model has

Table 5.2 NOAN analysis of AIDO-EM model.

	No. of alive nodes				
No. of rounds	ARSH-FATI	CIRP	R-LEACH	PSO-ECSM	Proposed model
0	100	100	100	100	100
1000	99	99	99	91	100
2000	98	92	1	0	100
3000	76	1	0	0	100
4000	6	0	0	0	99
5000	0	0	0	0	96
6000	0	0	0	0	58
7000	0	0	0	0	20

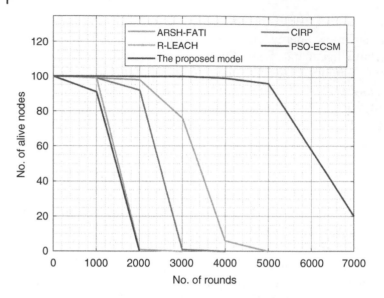

Figure 5.4 Comparative NOAN examination of AIDO-EM model.

resulted in higher NOAN of 100 whereas the ARSH-FATI, CIRP, R-LEACH, and PSO-ECSM models have obtained lower NOAN of 99, 99, 99, and 91, respectively. Next, with 2000 rounds, the AIDO-EM model has accomplished increased NOAN of 98 whereas the ARSH-FATI, CIRP, R-LEACH, and PSO-ECSM models have attained reduced NOAN of 98, 92, 1, and 0, respectively.

Hence, with 3000 rounds, the AIDO-EM model has resulted in higher NOAN of 100 whereas the ARSH-FATI, CIRP, R-LEACH, and PSO-ECSM models have obtained lower NOAN of 76, 1, 0, and 0, respectively. Followed by with 4000 rounds, the AIDO-EM model has resulted in higher NOAN of 99 whereas the ARSH-FATI, CIRP, R-LEACH, and PSO-ECSM models have obtained lower NOAN of 6, 0, 0, and 0, respectively.

The experimental result analysis of the AIDO-EM model with recent methods in terms of NRE is performed in Table 5.3 and Figure 5.5. The results indicated that the AIDO-EM model has reached maximum NRE

Table 5.3 NRE analysis of AIDO-EM model.

No. of rounds	Network remaining energy (NRE) (J)				
	ARSH-FATI	CIRP	R-LEACH	PSO-ECSM	Proposed model
0	25	25	25	25	25
500	22	21	21	18	24
1000	20	18	17	10	23
1500	18	12	11	5	22
2000	13	8	5	4	21
2500	10	5	4	0	20
3000	6	4	0	0	18
3500	5	0	0	0	17
4000	4	0	0	0	15
4500	0	0	0	0	12
5000	0	0	0	0	11
5500	0	0	0	0	8
6000	0	0	0	0	8
6500	0	0	0	0	7
7000	0	0	0	0	7

over the existing technique under all rounds. For instance, with 1000 rounds, the AIDO-EM model has resulted in higher NRE of 23 J whereas the ARSH-FATI, CIRP, R-LEACH, and PSO-ECSM models have obtained lower NRE of 20, 18, 17, and 10 J, respectively. Next, with 2000 rounds, the AIDO-EM model has accomplished increased NRE of 21 J whereas the ARSH-FATI, CIRP, R-LEACH, and PSO-ECSM models have attained reduced NRE of 13, 8, 5, and 4 J, respectively. Then, with 3000 rounds, the AIDO-EM model has resulted in higher NRE of 18 J whereas the ARSH-FATI, CIRP, R-LEACH, and PSO-ECSM models have obtained lower NRE of 6, 4, 0, and 0 J, respectively.

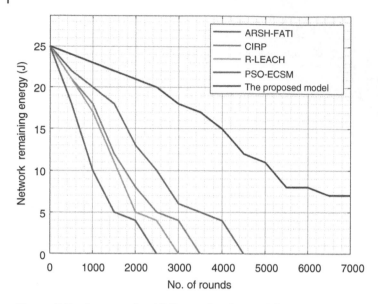

Figure 5.5 Comparative NRE examination of AIDO-EM model.

Followed by with 4000 rounds, the AIDO-EM model has resulted in higher NRE of 15 J whereas the ARSH-FATI, CIRP, R-LEACH, and PSO-ECSM models have obtained lower NRE of 4, 0, 0, and 0, respectively.

The comparative examination of the AIDO-EM model with recent methods in terms of number of packets sent to BS (NOPS-BS) is performed in Table 5.4 and Figure 5.6. The experimental values portrayed that the AIDO-EM model has accomplished increased NOPS-BS over the existing technique under all rounds.

For instance, with 1000 rounds, the AIDO-EM model has reached improved NOPS-BS of 58 024 whereas the ARSH-FATI, CIRP, R-LEACH, and PSO-ECSM models have resulted in reduced NOPS-BS of 33 148, 47 809, 52 835, and 54 930, respectively. Simultaneously, with 2000 rounds, the AIDO-EM model has accomplished increased NOPS-BS of 110 804 whereas the ARSH-FATI, CIRP, R-LEACH, and PSO-ECSM models have attained reduced NOPS-BS of 3 3985, 50 741,

Table 5.4   NOPS-BS analysis of AIDO-EM model.

	No. of packets sent to BS (NOPS-BS)				
No. of rounds	ARSH-FATI	CIRP	R-LEACH	PSO-ECSM	Proposed model
0	0	0	0	0	0
1000	33 148	47 809	52 835	54 930	58 024
2000	33 985	50 741	57 862	84 671	110 804
3000	33 985	50 741	59 119	85 508	143 382
4000	33 985	50 741	59 119	85 927	145 477
5000	33 985	50 741	59 119	85 927	145 990
6000	33 985	50 741	59 119	85 927	146 665
7000	33 985	50 741	59 119	85 927	147 084

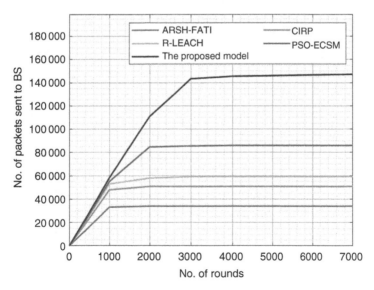

Figure 5.6   Comparative NOPS-BS examination of AIDO-EM model.

57 862, and 84 671, respectively. Concurrently, with 3000 rounds, the AIDO-EM model has resulted in higher NOPS-BS of 143 382 whereas the ARSH-FATI, CIRP, R-LEACH, and PSO-ECSM models have obtained lower NOPS-BS of 33 985, 50 741, 59 119, and 85 508, respectively. The aforementioned result analysis reported that the AIDO-EM model has surpassed all other existing methods in terms of distinct measures.

5.4 Conclusion

In this study, a novel AIDO-EM technique has been developed to achieve effective energy efficiency in the 6G networks. The presented AIDO-EM technique intends to accomplish minimal energy usage and maximum lifetime of the 6G enabled IoT devices. In addition, a novel DOA is applied for cluster enabled routing to achieve effective data distribution among the devices and choose effective GWHs. The DOA is a bio-inspired technique for global optimization that stimulates dingoes hunting strategy. In order to demonstrate the enhanced outcomes of the AIDO-EM technique, a series of simulations were carried out and the results reported the enhanced performance over its recent state-of-art approaches. Therefore, the AIDO-EM technique can be utilized as an effective way of accomplishing maximum energy management in 6G networks.

References

1 Yang, H., Alphones, A., Xiong, Z. et al. (2020). Artificial-intelligence-enabled intelligent 6G networks. *IEEE Network* 34 (6): 272–280.

2 Giordani, M., Polese, M., Mezzavilla, M. et al. (2020). Toward 6G networks: use cases and technologies. *IEEE Communications Magazine* 58 (3): 55–61.

3 Du, J., Jiang, C., Wang, J. et al. (2020). Machine learning for 6G wireless networks: carrying forward enhanced bandwidth, massive access, and

ultrareliable/low-latency service. *IEEE Vehicular Technology Magazine* 15 (4): 122–134.

4 Sodhro, A.H., Pirbhulal, S., Luo, Z. et al. (2020). Toward 6G architecture for energy-efficient communication in IoT-enabled smart automation systems. *IEEE Internet of Things Journal* 8 (7): 5141–5148.

5 Jiang, X., Sheng, M., Zhao, N. et al. (2021). Green UAV communications for 6G: a survey. *Chinese Journal of Aeronautics* 5: 621–632.

6 Janbi, N., Katib, I., Albeshri, A., and Mehmood, R. (2020). Distributed artificial intelligence-as-a-service (DAIaaS) for smarter IoE and 6G environments. *Sensors* 20 (20): 5796.

7 Wang, H., Huang, M., Zhao, Z. et al. (2020). Base station wake-up strategy in cellular networks with hybrid energy supplies for 6G networks in an IoT environment. *IEEE Internet of Things Journal* 8 (7): 5230–5239.

8 Mukherjee, A., Goswami, P., Khan, M.A. et al. (2020). Energy-efficient resource allocation strategy in massive IoT for industrial 6G applications. *IEEE Internet of Things Journal* 8 (7): 5194–5201.

9 Khan, W.U., Jameel, F., Jamshed, M.A. et al. (2020). Efficient power allocation for NOMA-enabled IoT networks in 6G era. *Physical Communication* 39: 101043.

10 Hu, N., Tian, Z., Du, X., and Guizani, M. (2021). An energy-efficient in-network computing paradigm for 6G. *IEEE Transactions on Green Communications and Networking* 5 (4): 1722–1733.

11 Sodhro, A.H., Zahid, N., Wang, L. et al. (2021). Towards ML-based energy-efficient mechanism for 6G enabled industrial network in box systems. *IEEE Transactions on Industrial Informatics* 17 (10): 7185–7192.

12 Mao, B., Kawamoto, Y., and Kato, N. (2020). AI-based joint optimization of QoS and security for 6G energy harvesting Internet of Things. *IEEE Internet of Things Journal* 7 (8): 7032–7042.

13 Liu, Y., Wang, X., Boudreau, G. et al. (2020). A multi-dimensional intelligent multiple access technique for 5G beyond and 6G wireless networks. *IEEE Transactions on Wireless Communications* 20 (2): 1308–1320.

14 Bairwa, A.K., Joshi, S., and Singh, D. (2021). Dingo optimizer: a nature-inspired metaheuristic approach for engineering problems. *Mathematical Problems in Engineering* 2021: 1–21.

15 Peraza-Vázquez, H., Peña-Delgado, A.F., Echavarría-Castillo, G. et al. (2021). A bio-inspired method for engineering design optimization inspired by dingoes hunting strategies. *Mathematical Problems in Engineering* 2021: 1–19.

16 Verma, S., Kaur, S., Khan, M.A., and Sehdev, P.S. (2020). Toward green communication in 6G-enabled massive internet of things. *IEEE Internet of Things Journal* 8 (7): 5408–5415.

6

Adaptive Whale Optimization with Deep Learning Enabled RefineDet Network for Vision Assistance on 6G Networks

Vinita Malik[1,2], Marcello Carvalho dos Reis[3,4], and Victor Hugo C. de Albuquerque[5]

[1] Graduate Program in Telecommunication Engineering, Federal Institute of Education, Science and Technology of Ceará, Fortaleza, Brazil
[2] Central Library Central University of Haryana, Mahendragarh, Haryana, India
[3] Graduate Program in Telecommunication Engineering, Federal Institute of Education, Science and Technology of Ceará, Fortaleza, Brazil
[4] Meteora, Fortaleza, Brazil
[5] Department of Teleinformatics Engineering, Federal University of Ceará, Fortaleza, Brazil

6.1 Introduction

As the global deployment of 5G networks progresses, next-generation (6G) network technology has been technologically advanced to assist artificial intelligence (AI) in deploying of "smart" network [1]. The fifth generation (5G) network provides reliable communication, higher speed, and lower latency services and makes greater difference in their day to day lives [2]. But it could be complex for meeting the increasing demand of Internet of Things (IoT) devices depending on the present technology, hence the idea of sixth generation (6G) network has been increased for improving the current 5G network and promoting the expansion of smart application. The vision of the 6G network is colorful; however, the comprehension of 6G network continues to be problematic, involving high throughput, high peak, high energy efficacy, connectivity all over the place, new technologies and theories, and other

AI-Enabled 6G Networks and Applications, First Edition. Edited by Deepak Gupta, Mahmoud Ragab, Romany Fouad Mansour, Aditya Khamparia, and Ashish Khanna.
© 2023 John Wiley & Sons Ltd. Published 2023 by John Wiley & Sons Ltd.

nontechnical problems [3]. It is recognized that the promising technology to develop 6G network includes: (i) basic techniques, namely sparse theory, flexible spectrum, new channel coding, and large-scale antenna; (ii) distinct techniques to realize space-air-ground-sea united communication and wireless tactile network; (iii) expanded spectrum transmission technique. At some point, the architecture of 6G network becomes more heterogeneous and larger, and application scenario becomes more variable and complex. It is nearly an unavoidable option for using AI to resolve this problem in wireless transmission techniques like signal transmission, resource allocation, and energy efficiency [1, 4]. Figure 6.1 illustrates the progress of 6G networks.

Science and engineering make technological intervention in the living of blind persons to make them independent to perceive and navigate surrounding objects [5].

Various devices are presented for assisting blind people; however, many of the devices concentrate on object recognition via computer vision (CV) or obstacle recognition via distinct sensor nodes namely distance sensors, GPS, and so on. But the efficient use of sensor-based techniques and CV might lead to supportive and very effective devices, making them aware of their surrounding [6]. The blind person uses electronic travel aids (ETA) for detecting problems and identifying facilities to offer informative and safe navigation. It decreases the risk of falling and preserves the balance of the person [7]. The weight, length, and cost are other parameters that are enhanced for best provision [8]. An electronic mobility cane (EMC) was developed for vision rehabilitation of blind persons for detecting problems and providing support [9], whereby a logical map is built to attain data regarding the surroundings. Output data is transported through voice, audio, or vibration.

Charan et al. [10] utilize deep learning (DL) and CV. It presents a novel solution that proactively forecasts dynamic connect blockages. In detail, it progresses a deep neural network (DNN) structure that learns in detected sequence of RGB images and beam-forming vector for predicting feasible more connect blockages. In Ref. [11], an enhanced object recognition technique dependent upon video key frame to latency decrease on edge internet of vehicles (IoV) scheme was presented. It considerably enhances latency reduction efficiency at the expense of lesser recognition

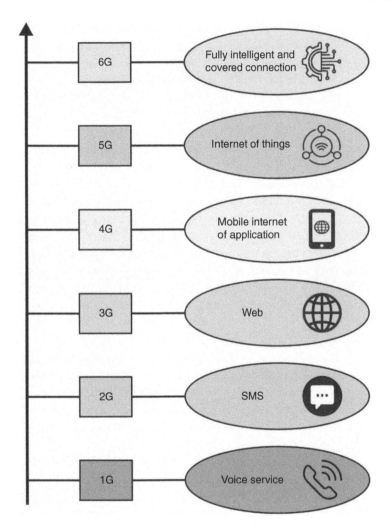

Figure 6.1 Evolution of 6G networks.

accuracy. During this scheme, it can be adapted to a vital coefficient and frame similarity comparison technique for filtering redundant frames and to attain key frame to object recognition. Afterward, an enhanced Haar-like feature based classifier technique was utilized for object recognition in the edge computation method.

Ukhwah et al. [12] utilized YOLO with three distinct infrastructure structures, for instance, Yolo v3, Yolo v3 Tiny, and Yolo v3 SPP allowed for creating a further accurate assessment to detect potholes on road surface. The authors presented a model with objective of helping the visually impaired by offering audio aid to refer them to avoiding obstacles that help them for moving from their surroundings. An object detection utilizing YOLO guides them in detecting the neighboring objects, and depth estimation utilizing monocular vision tells the estimated distance of detecting objects in the users.

Ning et al. [13] utilize transfer learning (TL) on single-shot detection (SSD) process to object recognition and classifier, then detection of human face and currency note, when noticed, utilizing Inception v3 method. SSD detector was trained on altered PASCAL VOC 2007 dataset, whereas a novel class was additional, for enabling the recognition of currency as well. Moreover, separate Inception v3 methods were training for recognizing human face and currency notes, so creating the infrastructure scalable and adjustable based on the user preference.

Joshi et al. [14] presented AI based fully automatic assistive technologies for recognizing distinct objects, and auditory inputs were offered to users from real time that provide optimum understanding to visually impaired person (VIP) about their environments. A DL technique was training with several images of objects that are extremely appropriate to VIP. Besides CV based approaches to object detection, a distance measuring sensor has been combined for making the devices further comprehensive by identifying obstacles but navigating from one place to another.

This study introduces an adaptive whale optimization with deep learning enabled RefineDet network (AWO-DLRDN) for visual assistance on 6G networks. The proposed AWO-DLRDN technique initially performs data augmentation and image annotation process as a preprocessing step. Moreover, the RefineDet model is applied for the identification of objects that exist in the image, and the hyperparameters of the RefineDet model are optimally adjusted by the use of AWO algorithm. Furthermore, approximate distances between the objects and persons are computed to assist visually impaired people. The experimental result analysis of the AWO-DLRDN technique is experimented with using benchmark dataset.

6.2 The Proposed Model

In this study, a novel AWO-DLRDN technique has been developed to determine the nearby objects and their approximate distance to the visually impaired people. The proposed AWO-DLRDN technique encompasses data augmentation, image annotation, RefineDet based object detection, AWO based hyperparameter optimization, and approximate distance estimation.

6.2.1 Image Augmentation and Annotation

Each gathered image is then augmented for resisting the trained method in overfitting and for performing further robust and accurate object recognition to VIPs. Several augmentations approach namely rotation at distinct angles, skewing, mirror, flip, brightness level, noise level, and group of these approaches are utilized for enriching the dataset to several folds. Every image is annotated manually with LabelImg tools, and bounding box is developed around an object with not taking extra unnecessary regions. The data on the image like the size of images, size, and place of bounding boxes (in the event, several samples or several objects from a similar image) are recorded and saved into ".xml" format. If the image is annotated, the respective annotation file is also created. The last dataset is of annotated image, and respective annotation file has been separated as two sets such as trained and validation.

6.2.2 RefineDet Based Object Detection

In order to detect the objects in an effective way, the RefineDet model has been applied. RefineDet [15] is a single-stage technique that depends on the SSD architecture and encompasses an object-detection module (ODM) and anchor-refinement model (ARM). The ARM pass negative hard-refined anchor and positive-refined anchor to the ODM that tries to classify and locate target objects in input image. In the current work, we utilized RefineDet as a benchmark model for the subsequent reasons: (i) it is effective because of its single-stage framework; and (ii) it employs a refined method that mimics a "detection method" for finding feasible regions in target traffic sign, nevertheless of their classes. Now, "detection method" varied from that utilized in conventional traffic sign detection (that is, recognize each function category specific among traffic signs in original image). RefineDet is an effective object-detection approach employed for detecting objects with higher accuracy and speed, but it is not competitive with advanced methodologies at smaller-object detection. First, feature of shallow layer in RefineDet and utilized for detecting smaller objects contain constraint data that is not strong enough to efficiently identify smaller objects.

6.2.3 Hyperparameter Tuning Using AWO Algorithm

For adjusting the hyperparameter values of the RefineDet model, the AWO algorithm has been employed. The searching of Whale Optimization Algorithm (WOA) is classified into two processes [16]: bubble-net attacking approach (exploitation process) and search for prey (exploitation phase). Initially, WOA implements searching for prey that is determined by:

$$D = |CX_{rand} - X(t)| \tag{6.1}$$

$$X(t + 1) = X_{rand} - AD \tag{6.2}$$

where D indicates the distance among the target and the present positions, t indicates the iteration number, A and C show coefficient vectors, $A = 2a \cdot r - a$, $C = 2r$, in which a is minimized linearly from 2 to 0 through iteration (exploration and exploitation stages) and r represents an arbitrary number within $[0, 1]$, X_{rand} indicates the location of an arbitrary whale in the population, and $X(t)$ represents the location of the existing whale. Usually, humpback whales use bubble-net attacking approach to feed as follows

$$X(t + 1) = \begin{cases} X^*(t) - AD & \text{if } p < 0.5 \\ D'e^{bl}\cos(2\pi l) + X^*(t) & \text{if } p \geq 0.5 \end{cases}$$ (6.3)

Here $D' = |X^*(t) - X(t)|$ and represents the distance of i from the whale to the prey (optimal solution attained until now), b indicates a constant for determining the shape of logarithmic spiral, l shows an arbitrary value in $[-1, 1]$, and p indicates an arbitrary value within in $[0, 1]$.

The balance among exploration and exploitation of metaheuristics was aim of the study that is associated with the overall effectiveness of the model [17]. When the exploration is stronger, the convergence precision is minimized; or else, it is easier to fall into the local optimal. Therefore, the study presents an adaptive improved exploratory method and formed the AWO algorithm. In the exploration stage of WOA, search for prey implies that one individual explores near other individuals from the population. In all the iterations, to spread the exploration region all over the solution space, the study presented an adoptive increased exploratory method. While whales adaptively choose the internal and external region of the population for exploring the size of the population space.

$$p(\delta_d) = 1 - \sin\left(\delta_d \times \frac{\pi}{2}\right)$$ (6.4)

$$\delta_d = \frac{\max(x) - \min(x)}{x_d^{\max} - x_d^{\min}} \quad d = 1, 2, ..., D$$ (6.5)

$X(t + 1)$

$$= \begin{cases} X_{rand} - AD & \text{if } p' \geq p(\delta_d) \\ \left. \begin{cases} rand * \left(x_d^{\max} - \max(x_d)\right) + \max(x_d) & \text{if } p'' \leq 0.5 \\ rand * \left(\min(x_d) - x_d^{\min}\right) + x_d^{\min} & \text{if } p'' \leq 0.5 \end{cases} \right\} & \text{if } p < p(\delta_d) \end{cases}$$

(6.6)

Here δ_d denotes the proportion of population to the solution space on d-th parameter, x_d represents a set of d-th variables of each individual, x_d^{max} and x_d^{min} show the upper and lower limits of the searching region, *rand*, p', and p'' indicate arbitrary vector from 0 to 1, $p(\delta_d)$ signifies the likelihood of exploring outside the population space, that is inversely proportionate to δ_d, and its curve. Figure 6.2 depicts the flowchart of WOA technique.

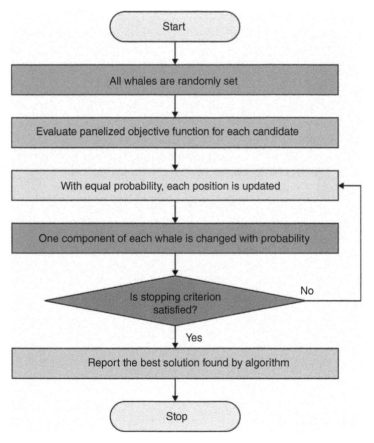

Figure 6.2 Flowchart of WOA.

6.2.4   Distance Measurement

For obtaining the distance among the user and object detected, it can require minimum of two points of view for mapping the image from 3D and creating depth map. The stereo vision utilizes two camera-calibrated images for perceiving the 3D infrastructure of world and creating their depth map. The calibration of two cameras is complete for knowing their intrinsic parameters, like image center, distortion, focal length, skew, and so on, and their extrinsic parameters that explain their place and orientation from their surroundings. For estimating the depth utilizing stereo vision, the focal length and distance among the two cameras can be required. At this point, x^l and x^r represent the distance among the optical axes and the projection of object "P" on virtual planes of two cameras. When b refers to the baseline distance among the two cameras and f signifies the focal length of the combined cameras, the perpendicular distance among object "P" and the baseline of cameras are provided as:

$$Z = \frac{b * F}{x^l + x^r} \tag{6.7}$$

6.3   Results and Discussion

The performance of the AWO-DLRDN model is tested using a dataset, comprising six classes with 200 images under every class. Table 6.1 and Figure 6.3 provide detailed object detection outcomes of the AWO-DLRDN model on the test images. The proposed AWO-DLRDN model has properly detected and recognized the objects.

On the person objects, the AWO-DLRDN model has correctly detected and recognized 198 and 198 objects, respectively. Besides, the AWO-DLRDN model has detected and recognized 197 and 196 car objects. Moreover, the AWO-DLRDN model has detected and recognized 194 and 192 truck objects.

Table 6.2 provides a brief results analysis of the AWO-DLRDN model in terms of detection accuracy (DA) and recognition accuracy. The table

Table 6.1 Result analysis of AWO-DLRDN technique under different classes.

Objects	Total testing images	Correctly detected	Correctly recognized
Person	200	198	198
Car	200	197	196
Bus	200	195	195
Truck	200	194	192
Chair	200	198	198
TV	200	192	192

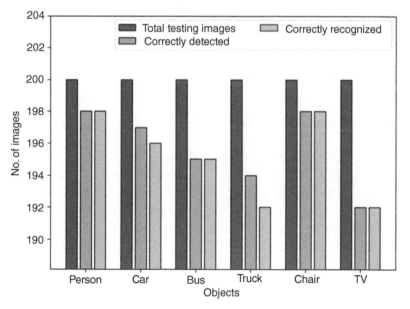

Figure 6.3 Result analysis of AWO-DLRDN technique under different classes.

Table 6.2 Result analysis of AWO-DLRDN technique in terms of DA and recognition accuracy.

Objects	Detection accuracy (%)	Recognition accuracy (%)
Person	99.00	100.00
Car	98.50	99.49
Bus	97.50	100.00
Truck	97.00	98.97
Chair	99.00	100.00
TV	96.00	100.00

indicated that the AWO-DLRDN model has detected person, car, bus, truck, chair, and TV objects with the increased DA of 99, 98.50, 97.50, 97, 99, and 96%, respectively. Besides, the AWO-DLRDN model has recognized person, car, bus, truck, chair, and TV objects with the increased recognition accuracy of 100, 99.49, 100, 98.97, 100, and 100%, respectively.

The accuracy outcome analysis of the AWO-DLRDN technique on the test data is illustrated in Figure 6.4. The results demonstrated that the AWO-DLRDN technique has accomplished improved validation accuracy compared to training accuracy. It is also observable that the accuracy values get saturated with the count of epochs.

The loss outcome analysis of the AWO-DLRDN technique on the test data is demonstrated in Figure 6.5. The figure revealed that the AWO-DLRDN technique has denoted the reduced validation loss over the training loss. It is additionally noticed that the loss values get saturated with the count of epochs.

Table 6.3 illustrates the comparative analysis of AWO-DLRDN technique with recent approaches in terms of testing accuracy and frame processing time (FPT).

Figure 6.6 demonstrates the testing accuracy investigation of the AWO-DLRDN model. The results reported that the AlexNet, VGG-16, and VGG-19 models have reached minimal testing accuracies of

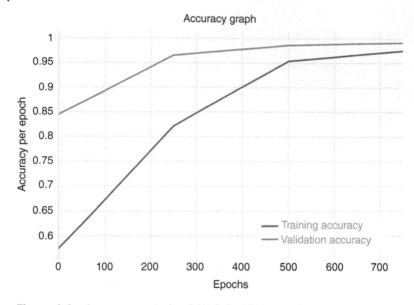

Figure 6.4 Accuracy analysis of AWO-DLRDN technique.

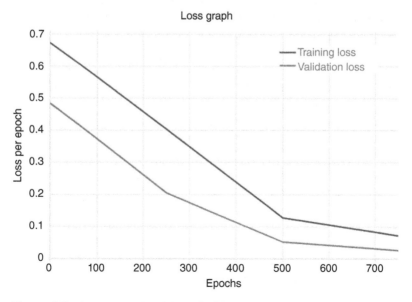

Figure 6.5 Loss analysis of AWO-DLRDN technique.

Table 6.3 Comparative analysis of AWO-DLRDN technique with recent approaches.

Methods	Testing accuracy	Frame processing time (ms)
AlexNet	83.89	275.00
VGG-16	87.00	530.00
VGG-19	90.20	390.00
YOLO-v3	95.20	100.00
AWO-DLRDN	98.74	75.00

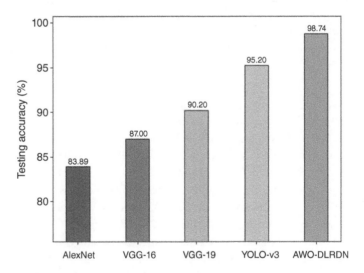

Figure 6.6 Testing accuracy analysis of AWO-DLRDN technique with existing approaches.

83.89, 87 and 90.20%, respectively. Along with that, the YOLO-v3 model has accomplished slightly enhanced testing accuracy of 95.20%. But the AWO-DLRDN model has accomplished higher testing accuracy of 98.74%.

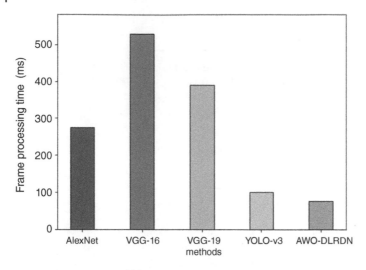

Figure 6.7 False positive ratio (FPR) analysis of AWO-DLRDN technique with existing approaches.

A comprehensive FPT inspection of the AWO-DLRDN model with other DL models is offered in Figure 6.7. The figure portrayed that the VGG-16 model has resulted in higher FPT of 390 ms whereas the AlexNet and VGG-19 models have obtained slightly reduced FPT of 275 and 390 ms, respectively. Although the YOLO-v3 model has resulted in reasonable FPT of 100 ms, the presented AWO-DLRDN model has outperformed the other methods with least FPT of 75 ms.

Finally, a computation time (CT) examination of the AWO-DLRDN model with other DL models is provided in Table 6.4 and Figure 6.8. The experimental results indicated that the VGG-16 model has resulted in higher CT of 12.94 seconds whereas the AlexNet and VGG-19 models have obtained slightly reduced CT of 6.71 and 9.52 seconds, respectively. Though the YOLO-v3 model has resulted in reasonable CT of 2.44 seconds, the presented AWO-DLRDN model has outperformed the other methods with least CT of 1.83 seconds. The results indicated the betterment of the AWO-DLRDN model over the other existing techniques.

Table 6.4 Computation time analysis of AWO-DLRDN technique with recent methods.

Methods	Computation time (s)
AlexNet	6.71
VGG-16	12.94
VGG-19	9.52
YOLO-v3	2.44
AWO-DLRDN	1.83

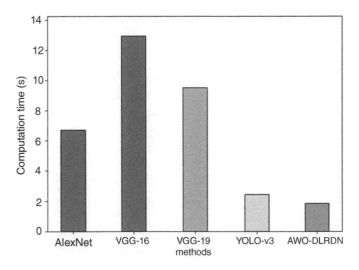

Figure 6.8 Computation time analysis of AWO-DLRDN technique with recent methods.

6.4 Conclusion

In this study, a novel AWO-DLRDN technique has been developed to determine the nearby objects and their approximate distance to the visually impaired people on 6G networks. The proposed AWO-DLRDN technique encompasses data augmentation, image annotation, Refin-Det based object detection, AWO based hyperparameter optimization, and approximate distance estimation. The hyperparameters of the Refi-neDet model are optimally adjusted by the use of AWO algorithm for accurate identification of objects that exist in the image. Lastly, approximate distances between the objects and persons are computed to assist visually impaired people. The experimental result analysis of the AWO-DLRDN technique is experimented using benchmark dataset and the comparison study highlighted the improvements of the AWO-DLRDN technique over other techniques. In future, the AWO-DLRDN technique can be realized in the smartphone application.

References

1 Yang, H., Alphones, A., Xiong, Z. et al. (2020). Artificial-intelligence-enabled intelligent 6G networks. *IEEE Network* 34 (6): 272–280.

2 Shafin, R., Liu, L., Chandrasekhar, V. et al. (2020). Artificial intelligence-enabled cellular networks: a critical path to beyond-5G and 6G. *IEEE Wireless Communications* 27 (2): 212–217.

3 Lovén, L., Leppänen, T., Peltonen, E. et al. (2019). EdgeAI: a vision for distributed, edge-native artificial intelligence in future 6G networks. *Proceeding of the First 6G Wireless Summit*, pp. 1–2, Levi, Finland, 24–26 March 2019. http://jultika.oulu.fi/Record/nbnfi-fe2019050314180.

4 Ye, Z. and Su, L. (2021). The use of data mining and artificial intelligence technology in art colors and graph and images of computer vision under 6G internet of things communication. *International Journal of System Assurance Engineering and Management* 12 (4): 689–695.

5 Janbi, N., Katib, I., Albeshri, A., and Mehmood, R. (2020). Distributed artificial intelligence-as-a-service (DAIaaS) for smarter IoE and 6G environments. *Sensors* 20 (20): 5796.

6 Liu, L., Ouyang, W., Wang, X. et al. (2020). Deep learning for generic object detection: a survey. *International Journal of Computer Vision* 128 (2): 261–318.

7 Wang, W., Lai, Q., Fu, H. et al. (2022). Salient object detection in the deep learning era: an in-depth survey. *IEEE Transactions on Pattern Analysis and Machine Intelligence* 44 (6): 3239–3259.

8 Wei, J., He, J., Zhou, Y. et al. (2019). Enhanced object detection with deep convolutional neural networks for advanced driving assistance. *IEEE Transactions on Intelligent Transportation Systems* 21 (4): 1572–1583.

9 Pathak, A.R., Pandey, M., Rautaray, S., and Pawar, K. (2018). Assessment of object detection using deep convolutional neural networks. In: (eds. Bhalla, S., Bhateja, V., Chandavale, A., Hiwale, A., Satapathy, S.) *Intelligent Computing and Information and Communication*, 457–466. Singapore: Springer.

10 Charan, G., Alrabeiah, M., and Alkhateeb, A. (2021). Vision-aided dynamic blockage prediction for 6G wireless communication networks. *2021 IEEE International Conference on Communications Workshops (ICC Workshops)*, 14–23 June 2021, Montreal, QC, Canada, pp. 1–6. IEEE. doi: 10.1109/ICCWorkshops50388.2021.9473651

11 Dai, C., Liu, X., Chen, W., and Lai, C.F. (2020). A low-latency object detection algorithm for the edge devices of IoV systems. *IEEE Transactions on Vehicular Technology* 69 (10): 11169–11178.

12 Ukhwah, E.N., Yuniarno, E.M., and Suprapto, Y.K. (2019). Asphalt pavement pothole detection using deep learning method based on YOLO neural network. *2019 International Seminar on Intelligent Technology and Its Applications (ISITIA)*, 28–29 August 2019, Surabaya, Indonesia, pp. 35–40. IEEE. doi: 10.1109/ISITIA.2019.8937176

13 Ning, C., Zhou, H., Song, Y., and Tang, J. (2017). Inception single shot multibox detector for object detection. *2017 IEEE International Conference on Multimedia & Expo Workshops (ICMEW)*, 10–14 July 2017, Hong Kong, China, pp. 549–554. IEEE. doi: 10.1109/ICMEW.2017.8026312

14 Joshi, R.C., Yadav, S., Dutta, M.K., and Travieso-Gonzalez, C.M. (2020). Efficient multi-object detection and smart navigation using artificial intelligence for visually impaired people. *Entropy* 22 (9): 941.

15 Zhang, S., Wen, L., Bian, X. et al. (2018). Single-shot refinement neural network for object detection. *Proceedings of the IEEE Conference on Computer Vision and Pattern Recognition*, 18–23 June 2018, Salt Lake City, UT, USA pp. 4203–4212. doi: 10.1109/CVPR.2018.00442

16 Mirjalili, S. and Lewis, A. (2016). The whale optimization algorithm. *Advances in Engineering Software* 95: 51–67.

17 Li, Y., Han, T., Zhao, H., and Gao, H. (2019). An adaptive whale optimization algorithm using Gaussian distribution strategies and its application in heterogeneous UCAVs task allocation. *IEEE Access* 7: 110138–110158.

7

Efficient Deer Hunting Optimization Algorithm Based Spectrum Sensing Approach for 6G Communication Networks

R. Pandi Selvam[1], Kanagaraj Narayanasamy[2], and M. Ilayaraja[3]

[1] *PG Department of Computer Science, Vidhyaa Giri College of Arts and Science, Puduvayal, Karaikudi, Tamilnadu, India*
[2] *Department of Computer Applications, Karpagam Academy of Higher Education (Deemed to be University), Coimbatore, Tamilnadu, India*
[3] *School of Computing, Kalasalingam Academy of Research and Education, Krishnankoil, Tamilnadu, India*

7.1 Introduction

The technological generation in wireless communication plays an important role to provide users with minimum latency and high data rate [1]. The progression of 5G is anticipated to offer a major involvement to spectrum supervision, public safety, high data rates, and energy efficiency [2]. Internet of Things (IoT) is the center of 5G and 6G techniques. Therefore, the IoT-based device plays a major role in the framework of 5G and 6G network systems [3]. The 6G system is predicted to resolve the limitation of 5G networks. They allow users to provide the request for low latency, high throughput, and large capacity. 6G network requires the efficient usage of the spectrum to satisfy the user demand. Therefore, the 6G network needs to utilize new techniques such as cognitive radios (CRs). Thus, the tremendous growth in wireless technology and the number of devices request new wireless services in the underutilized and utilized portions of the radio spectrum [4]. In this context, spectrum sensing (SS) was legalized through the federal communication commission (FCC) [5]. The utilization of CR in wireless

AI-Enabled 6G Networks and Applications, First Edition. Edited by Deepak Gupta, Mahmoud Ragab, Romany Fouad Mansour, Aditya Khamparia, and Ashish Khanna.
© 2023 John Wiley & Sons Ltd. Published 2023 by John Wiley & Sons Ltd.

transmission offers sufficient intellect with effective radio spectrum use that adjusts and learns the transmission device parameter [6]. The primary user (PU) is authorized in cognitive radio network (CRN) for accessing the spectrum, whereas the secondary user (SU) is opportunist for accessing the PU spectrum. The SU needs to evacuate the spectrum on the PU arrival to prevent intervention with the genuine PU. The uncertainty in the wireless network, i.e. the hidden terminal, fading, and shadowing problem, limits performance of the user senses. Figure 7.1 depicts the classification of SS methods.

Cooperative spectrum sensing (CSS) is taken into account as a potential resolution to resolve the difficulty. In the CSS, the user report and sense the finding to the fusion center (FC) [7]. But the key challenge in

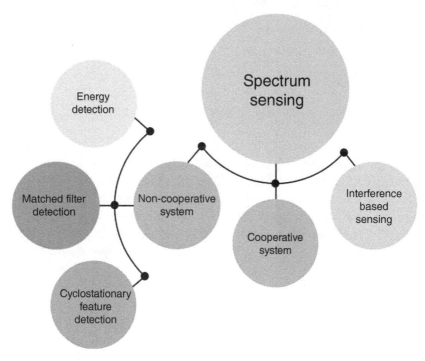

Figure 7.1 Classification of spectrum sensing methods.

the CSS decision is the identifying report of malevolent users (MU), whereby false data is stated to the FC for hijacking the FC decision [8]. With the popularization of mobile portable devices, PU and CR users (SU) are unlimited to fixed devices. Prime user has mobility in numerous cases, namely city mobile base stations and cognitive vehicles, whereby the conventional systems for SS do not perform well. The mobility of PU results in position variation. Consequently, the sensing ability of PU differs in time [9]. The mobility of PU leads to wrong sensing information. For detecting mobile PU, SU needs to move to the signal coverage of PU that should depend on measured received signal strengths (RSS). Few CRs track PU by position technique from third party [10]. Many effective approaches, game theory, machine learning (ML), and convex optimization are utilized for solving this problem. As an opposite alternative, swarm intelligence (SI), a set of artificial intelligence (AI), was widely and recently employed using remarkable performance.

Optimum utilization and organization of the accessible spectrum are important to enhance network efficiency. Therefore, this paper presents a novel efficient deer hunting optimization algorithm based spectrum sensing approach (EDHO-SSA) for 6G communication networks. The presented EDHO-SSA technique mainly intends to manage the availability of spectrums that exist in the 6G networks. The EDHO-SSA technique is based on the hunting nature of the deers. It also derives an objective function to define the performance of SS including distinct parameters such as energy and throughput. The experimental result analysis of the EDHO-SSA technique is carried out, and the results are assessed in terms of different measures.

7.2 Related Works

In Eappen and Shankar [11], a new multiobjective modified gray wolf optimizer (MOMGWO) approach was presented for solving the multiobjective optimized issue from the domain of SS from a CRN that is a vital paradigm from wireless communication technology. Modification in GWO was executed for balancing the trade-off among exploration as well as exploitation procedure from conventional GWO for obtaining global optimum. Chen et al. [12] examined a novel method named particle swarm optimization (PSO) based agent CSS (PSOA). During this method, it can utilize several mobile agents spread over the networks, for CSS rather than SUs. Each agent is moved based on the latest global optimum agent of equivalent target PU with fitness function (FF) computed by modified PSO.

In Tekbıyık et al. [13], a convolution neural network (CNN) method utilizing spectral correlation function (SCF) that is effectual classification of cyclostationarity property was presented to wireless SS and signal identity. The presented technique classifies wireless signal with no prior data, and it can be applied from two distinct settings entitled CASE1 and CASE2. Sheng et al. [14] offered an efficient trace of covariance matrix and variance of quadratic covariance matrix with support vector machine (TCVQ-SVM) approach dependent upon ML to narrowband SS. Primarily, trace of covariance matrix and variance of quadratic covariance matrix (TCVQ) was extracted as feature vector and integrated to trained instance of SS. Next, the classification method was attained by training instances dependent upon SVM that avoids setting threshold and altering classifier hyperplane by its self-learning abilities.

Chen et al. [15] presented a relay supported maximum ratio combining/zero forcing beam-forming (MRC/ZFB) method for guaranteeing the secret performance of dual-hop short-packet communication from cognitive IoT. This work examines the average confidentiality throughput of model and examines two asymptotic situations with higher signal to noise ratio (SNR) regimes and infinite blocklength. Eappen et al. [16] employed the unused spectrum of the CR relay supported cooperative device to device (D2D) communication is purpose for enhancing the network capacity (data rate) of user equipment (UE)/licensed user (LU) that is network blockage region to wireless remote patient monitoring system (PMS). The CR allowed with modified Whale optimization algorithm (WOA) (MWOA) was presented to effectual SS.

7.3 The Proposed Model

This paper has developed a new EDHO-SSA technique for optimal SS in 6G communication networks. The presented EDHO-SSA technique has properly managed the availability of spectrums that exist in the 6G networks. The EDHO-SSA technique is based on the hunting nature of the deer. It also derives an objective function to define the performance of SS including distinct parameters such as energy and throughput. The metaheuristic DHO approach initiates by the set called the hunter that is collection of arbitrary populations, determined by the following:

$$Z = \{y_1, y_2, ..., y_n\}, \quad 1 < j \leq n, \tag{7.1}$$

Where the types of solutions or number of hunters are represented as n. As well, Z denotes the overall hunter population. The next phase includes quantifying the major component, the angle of wind, and location of the deer. Space is taken into account as a circle. Thus, the wind angle is reformulated as:

$$\theta_j = 2\pi\alpha, \tag{7.2}$$

Where α indicates an arbitrary number in $\{0, 1\}$ and j defines the existing iteration. Similarly, the angle of deer place is determined by:

$$\phi_j = \pi + \beta. \tag{7.3}$$

In this formula, β indicates the angle of wind. In the initial iteration, it can be generally impossible to discover the optimal solution for the approach [17]. But after creating an arbitrary number and evaluating the cost function from it, the optimal number is taken into account as the optimum solution [18]. Now, consider two variables, involving the successor location (Z) as the following hunter position and leader location (Z) as the first optimal position of the hunter.

To obtain the optimum place with the primary repetition, the whole population tries to attain the optimum place by upgrading the position:

$$Z_{j+1} = Z_L - k \times S^w \times \left| L \times Z_L - Z_j \right|. \tag{7.4}$$

Now, Z_j and Z_{j+1} indicates the existing and later positions, S^w indicates the arbitrary number that depends on wind velocity within $[0, 2]$, and the coefficient vector is represented as L and k:

$$k = 0.25 \times \log \left(I + \frac{1}{I_{\max}} \right) \gamma, \tag{7.5}$$

$$L = 2 \times \delta, \tag{7.6}$$

Where $I_{\max}$ denotes the peak of repetition within $[-1, 1]$, and γ shows the arbitrary element. δ indicates the arbitrary number within 0 to 1. For upgrading location of Z^*, where (Z, Y) indicates the prime position of the hunter that is upgraded according to the prey position. The upgraded status would be enduring to attain an optimal state (Z^*, Y^*) according to the L and K. The hunter goes to the place in which the leader is situated. When the leader's moves were not effective, the hunter stays in their preceding location. The upgrading of location is based on Eq. (7.6), if $S^w < 1$. In fact, hunter moves in each direction nevertheless of the location angle. Consequently, based on Eqs. (7.6) and (7.7), the hunter upgrades the location in all the arbitrary locations.

As well, extend the space of resolving method by assuming the position angle. Angle calculation is significant for determining the location of the hunter. Hence, the effective attack should not be noticeable to the prey. The visualization of deer angle (prey) equation is given as follows:

$$u_j = \frac{1}{8} \times \pi \times \alpha. \tag{7.7}$$

Because of the variance among the angle of wind and the angle where the prey is seen, u denotes the variable that is taken into account to update the angle of location:

$$C_j = \beta_j - u_j, \tag{7.8}$$

where β shows the angle of the wind blowing.

Next, for updating the location angle variable,

$$\phi_{j+1} = \phi_j + C_j. \tag{7.9}$$

After attaining the angle of position, the novel position is estimated by:

$$Z_{j+1} = Z_j - S^w \times \left| \cos\left(\varphi_{j+1}\right) \times Z_l - Z_i \right|. \tag{7.10}$$

The prey does not see the hunter due to the view angle. To utilize the exploration, it is probable to alter L in the performance of encircling. Based on arbitrary searching, the number of vectors L could not be considered as more than 1.

$$Z_{j+1} = Z_s - k \times S^w \times \left| L \times Z_L - Z_j \right|, \tag{7.11}$$

Where Z_s denotes the successor position of hunter at any moment. Here, the position of the hunter is upgraded by the optimal solution in all the repetitions. The optimal solution is attained when $|L| \geq 1$. When $|L| < 1$, one of the hunters is arbitrarily chosen. This technique generates an L switch that could adjust the mode of approach among the exploitation and exploration phases.

CRN includes SU present in a similar geographical position and shares a similar spectrum [19]. Time frame (T^P) of CRN is separated into two time slots as follows:

The first time slot is the SS represented as t^s, and next is the data transmission, characterized by t^D. Hypotheses for SS problem with $j = 1, ..., n^s$ amount of secondary users, $t = 1, ..., n^{sc}$ amount of subcarriers, and time index $i = 1, 2, ..., k_j$ is expressed as follows:

$$H_{l0}: y_{j,l}(i) = w_{j,l}(i)\{Hypotheses\ 0\ (PU\ Absent)\} \tag{7.12}$$

$$H_{l1}: y_{j,l}(i) = S_{j,l}(i) + w_{j,l}(i)\ \{Hypotheses\ 1\ (PU\ Present)\}$$

Figure 7.2 illustrates the SS in CRN. Now, H_{l0} denotes the Hypothesis 0 for lth subcarrier that represented the nonattendance of PU signal and H_{l1} shows the Hypothesis 1 for the lth subcarrier indicating the existence of PU; $w_{j,l}$ indicates the white noise having Gaussian distribution using mean zero and variance $\sigma_{j,l}^2$. $S_{j,l}$ denotes the main signaling analogous to stationary arbitrary procedure with zero mean and variance $\sigma_{s_{j,l}}^2$. Sampling frequency and sensing time for jth SU are signified by f_j^s and t_j^s,

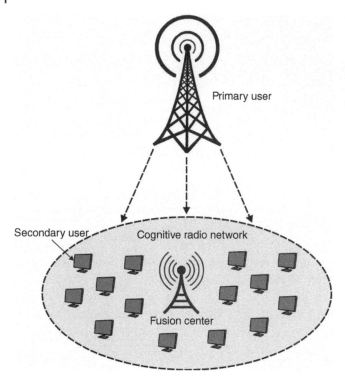

Figure 7.2 Spectrum sensing in cognitive radio networks.

correspondingly; then, the amount of instance could be represented by $k_j = f_j^s t_j^s$. Energy detection based Neyman–Pearson structure decision rule for jth SU across lth subcarrier was estimated by:

$$E_{j,l}^n = \frac{1}{k_j} \sum_{i=1}^{k_j} |y_{j,l}(i)| 2H_{l1} > \lambda_{j,l} < H_{l0} \tag{7.13}$$

$$E_{j,l}^n \mid H_{lr} \sim N\left(\mu_{j,l|r}, \sigma_{j,l|r}^2 / k_j\right) \tag{7.14}$$

where $r = 0, 1$ based on the hypothesis and energy detection for distinct hypothesis and larger value of k_j follows standard distribution with mean and variance as $\mu_{j,\,l|r}, \sigma_{j,l|r}^2$, correspondingly.

$$\mu_{j,l|r} = \begin{cases} \sigma_{j,l}^2 & \text{if } r = 0 \\ \sigma_{S_{j,l}}^2 + \sigma_{j,l}^2 & \text{if } r = 1 \end{cases} \tag{7.15}$$

and

$$\sigma_{j,l|r}^2 = \begin{cases} \sigma_{j,l|r}^4 & \text{for } r = 0 \\ E\left|S_{j,l}\right|^4 + 2\sigma_{j,l}^2 - \left(\sigma_{S_{j,l}}^2 - \sigma_{j,l}^2\right)^2 & \text{for } r = 1 \end{cases} \tag{7.16}$$

Considering PU signal as Gaussian, the preceding equation is reformulated by:

$$\sigma_{j,l|r}^2 = \begin{cases} \sigma_{j,l|r}^4 & \text{for } r = 0 \\ \left(\sigma_{S_{j,l}}^2 + \sigma_{j,l}^2\right) & \text{for } r = 1 \end{cases} \tag{7.17}$$

with the aforementioned consideration, the detection probability and false alarm probability is estimated by:

$$\text{prob} f_{j,l}\left(\lambda_{j,l}, t_j^s\right) = Q\left(\sqrt{t_j^s f_j^s} \frac{\lambda_{j,l} - \mu_{j,l|0}}{\sigma_{j,l|0}}\right) \tag{7.18}$$

$$\text{prob } d_{j,l}\left(\lambda_{j,l}, t_j^s\right) = Q\left(\sqrt{t_j^s f_j^s} \frac{\lambda_{j,l} - \mu_{j,l|1}}{\sigma_{j,l|1}}\right) \tag{7.19}$$

7.4 Experimental Validation

This section inspects the performance validation of the EDHO-SSA technique over the other existing techniques.

Table 7.1 and Figure 7.3 provide an average energy efficiency (AEE) of the EDHO-SSA technique with other methods under varying levels of interference. The results indicated that the EDHO-SSA model has resulted in increased AEE over the other methods under all levels of interferences. For instance, with 0.50 mW interference, the EDHO-SSA model has provided higher AEE of 2.8893 J whereas the MOMGWO, multiobjective particle swarm optimization (MOPSO), multi-objective cuckoo search optimization (MOCSO), multiobjective

Table 7.1 Average energy efficiency analysis of EDHO-SSA technique with other methods.

Interference (mW)	Average energy efficiency (J)					
	EDHO-SSA	MOMGWO	MOPSO	MOCSO	MOGWO	NSGA-II
0	0.0000	0.0000	0.0000	0.0000	0.0000	0.0000
0.25	1.6923	1.7081	1.6766	1.7081	1.5663	1.3513
0.50	2.8893	2.8578	2.7948	2.3381	1.9286	1.6906
1.00	4.9130	4.7398	3.8610	2.7790	2.2593	1.9783
1.12	5.4092	4.8737	4.4957	3.6138	2.7318	2.4848
1.24	5.6139	5.0627	4.7792	4.4170	3.9602	3.7512
1.36	5.7872	5.2675	4.9840	4.6690	4.0075	3.7515
1.48	5.9289	5.5194	5.1100	4.8737	4.1965	3.9675
1.60	6.1022	5.7557	5.3147	4.9525	4.3225	4.1195
1.72	6.2282	5.8974	5.4722	5.0155	4.5587	4.3177
1.84	6.3542	6.0549	5.6297	5.1730	4.7320	4.4730
1.96	6.4329	6.1967	5.7399	5.2517	4.9525	4.7035
2.08	6.5747	6.2597	5.8187	5.3462	5.0155	4.7325
2.20	6.7164	6.3699	5.9132	5.4249	5.0942	4.8202
2.32	6.8581	6.4959	6.0077	5.4722	5.1572	4.8862
2.44	6.9526	6.5274	6.1022	5.6612	5.2832	5.0542
2.56	7.1259	6.5904	6.1809	5.6612	5.3304	5.0544
2.68	7.1731	6.6377	6.2439	5.7242	5.4092	5.2082
2.80	7.2834	6.7164	6.3384	5.8187	5.4722	5.1942
2.92	7.3149	6.7637	6.4329	5.8817	5.5509	5.2969
3.04	7.3464	6.8424	6.5117	5.8817	5.5824	5.3044
3.16	7.3936	6.9211	6.5589	5.9604	5.5667	5.3347
3.28	7.4094	7.0471	6.6534	6.0707	5.7084	5.4774
3.40	7.4566	7.0944	6.7007	6.1337	5.8029	5.5839
3.52	7.5039	7.1731	6.7007	6.2282	5.8502	5.5702

Table 7.1 (Continued)

	Average energy efficiency (J)					
Interference (mW)	EDHO-SSA	MOMGWO	MOPSO	MOCSO	MOGWO	NSGA-II
3.64	7.5511	7.1889	6.8109	6.2912	5.8974	5.6954
3.76	7.5511	7.2519	6.8109	6.3227	5.9604	5.6944
3.88	7.5354	7.2834	6.8581	6.3542	5.9919	5.7559
4.00	7.5354	7.3779	6.8896	6.3857	6.0234	5.7964

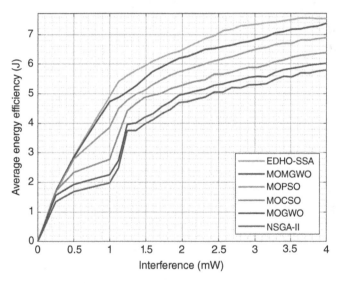

Figure 7.3 AEE analysis of EDHO-SSA technique with other methods.

grey wolf optimization (MOGWO), and non-domination based genetic algorithm (NSGA-II) models have offered lower AEE of 2.8578, 2.7948, 2.3381, 1.9286, and 1.6906 J, respectively. In addition, with 1.60 mW interference, the EDHO-SSA method has offered maximum AEE of

6.1022 J whereas the MOMGWO, MOPSO, MOCSO, MOGWO, and NSGA-II approaches have offered lower AEE of 5.7557, 5.3147, 4.9525, 4.3225, and 4.1195 J correspondingly. Along with that, with 2.20 mW interference, the EDHO-SSA system has provided higher AEE of 6.7164 J whereas the MOMGWO, MOPSO, MOCSO, MOGWO, and NSGA-II models have offered lower AEE of 6.3699, 5.9132, 5.4249, 5.0942, and 4.8202 J correspondingly. Moreover, with 3.04 mW interference, the EDHO-SSA system has provided higher AEE of 7.3464 whereas the MOMGWO, MOPSO, MOCSO, MOGWO, and NSGA-II models have offered lower AEE of 6.8424, 6.5117, 5.8817, 5.5824, and 5.3044 J correspondingly. At last, with 4.00 mW interference, the EDHO-SSA algorithm has provided higher AEE of 7.5354 J whereas the MOMGWO, MOPSO, MOCSO, MOGWO, and NSGA-II models have offered lower AEE of 7.3779, 6.8896, 6.3857, 6.0234, and 5.7964 J correspondingly.

Table 7.2 and Figure 7.4 provide the iterative generalized demodulation (IGD) inspection of the EDHO-SSA model over the other methods. The results showcased that the EDHO-SSA technique has outperformed the other methods in terms of IGD. For instance, the EDHO-SSA model has obtained best IGD of 50.12% whereas the MOPSO, NSGA-II, MOGWO, and MOCSO algorithms have attained best IGD of 55.44, 55.63, 62.92, and 62.93%, respectively. Eventually, the EDHO-SSA system has obtained worst IGD of 45.20% whereas the MOPSO, NSGA-II, MOGWO, and MOCSO algorithms have attained worst IGD of

Table 7.2 IGD analysis of EDHO-SSA technique with recent approaches.

	IGD (%)				
Methods	**EDHO-SSA**	**MOPSO**	**NSGA-II**	**MOGWO**	**MOCSO**
Best	50.12	55.44	55.63	62.92	62.93
Worst	45.20	50.17	59.58	57.47	47.63
Std dev.	80.02	81.97	86.69	86.69	87.01
Average	60.00	62.64	68.51	68.90	66.27

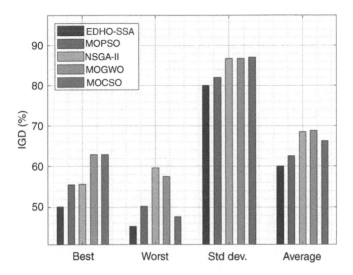

Figure 7.4 IGD analysis of EDHO-SSA technique with recent approaches.

Table 7.3 Spacing analysis of EDHO-SSA technique with recent approaches.

	Spacing (%)				
Methods	**EDHO-SSA**	**MOPSO**	**NSGA-II**	**MOGWO**	**MOCSO**
Best	94.36	91.60	76.02	94.94	90.71
Worst	43.29	57.10	50.50	58.93	57.64
Std dev.	88.11	85.95	53.53	82.25	87.36
Average	91.35	89.67	77.41	88.94	90.49

50.17, 59.58, 57.47, and 47.63% correspondingly. Meanwhile, the EDHO-SSA model has gained average IGD of 60% whereas the MOPSO, NSGA-II, MOGWO, and MOCSO algorithms have achieved average IGD of 62.64, 68.51, 68.90, and 66.27% correspondingly.

Table 7.3 and Figure 7.5 offer the spacing analysis of the EDHO-SSA model over the other methods. The outcomes illustrated that the

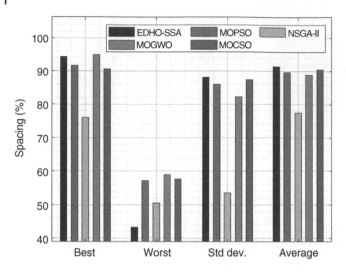

Figure 7.5 Spacing analysis of EDHO-SSA technique with recent approaches.

EDHO-SSA approach has exhibited the other approaches with respect to spacing. For sample, the EDHO-SSA approach has obtained best spacing of 94.36% whereas the MOPSO, NSGA-II, MOGWO, and MOCSO algorithms have attained best spacing of 91.60, 76.02, 94.94, and 90.71%, respectively. Followed by, the EDHO-SSA model has reached worst spacing of 43.29% whereas the MOPSO, NSGA-II, MOGWO, and MOCSO algorithms have attained worst spacing of 57.10, 50.50, 58.93, and 57.64% correspondingly. In the meantime, the EDHO-SSA method has obtained average spacing of 91.35% whereas the MOPSO, NSGA-II, MOGWO, and MOCSO systems have attained average spacing of 89.67, 77.41, 88.94, and 90.49%, respectively.

Table 7.4 and Figure 7.6 give the error ratio inspection of the EDHO-SSA system over the other methods. The results revealed that the EDHO-SSA method has exhibited the other algorithms in terms of error ratio. For instance, the EDHO-SSA model has obtained best error ratio of 80.38% whereas the MOPSO, NSGA-II, MOGWO, and MOCSO algorithms have attained best error ratio of 84.03, 90.46, 87.32, and 85.97%, respectively. Likewise, the EDHO-SSA model has obtained worst error

Table 7.4   IGD analysis of EDHO-SSA technique with recent approaches.

Methods	Error ratio (%)				
	EDHO-SSA	MOPSO	NSGA-II	MOGWO	MOCSO
Best	80.38	84.03	90.46	87.32	85.97
Worst	61.75	71.67	75.52	70.23	70.72
Std dev.	88.10	91.11	91.40	92.73	91.82
Average	84.40	85.34	90.96	86.81	88.97

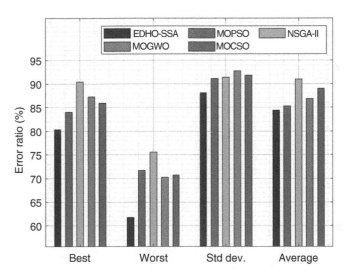

Figure 7.6   Error ratio analysis of EDHO-SSA technique with recent approaches.

ratio of 61.75% whereas the MOPSO, NSGA-II, MOGWO, and MOCSO algorithms have attained worst error ratio of 71.67, 75.52, 70.23, and 70.72%, respectively. Finally, the EDHO-SSA methodology has obtained average error ratio of 84.40% whereas the MOPSO, NSGA-II, MOGWO, and MOCSO algorithms have gained average error ratios of 85.34, 90.96, 86.81, and 88.97 correspondingly.

Table 7.5 and Figure 7.7 provide the maximum spread analysis of the EDHO-SSA methodology over the other methods. The outcomes exposed

Table 7.5 Maximum spread analysis of EDHO-SSA technique with recent approaches.

	Maximum spread (%)				
Methods	**EDHO-SSA**	**MOPSO**	**NSGA-II**	**MOGWO**	**MOCSO**
Best	35.00	37.43	40.49	39.30	36.84
Worst	24.24	28.22	30.05	30.33	27.20
Std dev.	21.23	23.69	17.76	4.28	10.59
Average	29.61	34.33	39.69	37.98	31.09

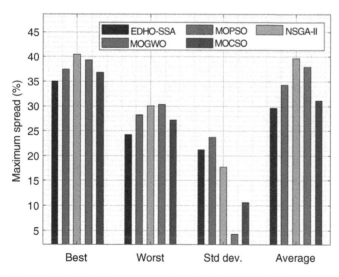

Figure 7.7 Maximum spread analysis of EDHO-SSA technique with recent approaches.

that the EDHO-SSA model has demonstrated the other approaches with respect to maximum spread. For instance, the EDHO-SSA model has obtained best maximum spread of 35% whereas the MOPSO, NSGA-II, MOGWO, and MOCSO algorithms have attained best maximum spread of 37.43, 40.49, 39.30, and 36.84% correspondingly. At the same time, the EDHO-SSA technique has obtained worst maximum spread of 24.24% whereas the MOPSO, NSGA-II, MOGWO, and MOCSO algorithms have attained worst maximum spread of 28.22, 30.05, 30.33, and 27.20 correspondingly. Eventually, the EDHO-SSA approach has obtained average maximum spread of 29.61% whereas the MOPSO, NSGA-II, MOGWO, and MOCSO algorithms have attained average maximum spread of 34.33, 39.69, 37.98, and 31.09% correspondingly.

7.5 Conclusion

This paper has developed a new EDHO-SSA technique for optimal SS in 6G communication networks. The presented EDHO-SSA technique has properly managed the availability of spectrums that exist in the 6G networks. The EDHO-SSA technique is based on the hunting nature of the deers. It also derives an objective function to define the performance of SS including distinct parameters such as energy and throughput. The experimental result analysis of the EDHO-SSA technique is carried out, and the results are assessed with respect to distinct measures. The experimental results reported the improved outcomes of the EDHO-SSA technique over the other techniques. In future, hybrid deep learning techniques will be executed to improve the SS performance.

References

1 Akyildiz, I.F., Lo, B.F., and Balakrishnan, R. (2011). Cooperative spectrum sensing in cognitive radio networks: A survey. *Physical Communication* 4 (1): 40–62.
2 Zhang, W., Mallik, R.K., and Letaief, K.B. (2008). Cooperative spectrum sensing optimization in cognitive radio networks. *2008 IEEE*

International Conference on Communications, pp. 3411–3415 (19–23 May 2008). Beijing, China: IEEE.

3 Ma, J., Zhao, G., and Li, Y. (2008). Soft combination and detection for cooperative spectrum sensing in cognitive radio networks. *IEEE Transactions on Wireless Communications* 7 (11): 4502–4507.

4 Ganesan, G. and Li, Y. (2007). Cooperative spectrum sensing in cognitive radio, part II: multiuser networks. *IEEE Transactions on Wireless Communications* 6 (6): 2214–2222.

5 Sun, C., Zhang, W., and Letaief, K.B. (2007). Cluster-based cooperative spectrum sensing in cognitive radio systems. *2007 IEEE International Conference on Communications*, pp. 2511–2515 (24–28 June 2007). Glasgow, UK: IEEE.

6 Atapattu, S., Tellambura, C., and Jiang, H. (2011). Energy detection based cooperative spectrum sensing in cognitive radio networks. *IEEE Transactions on Wireless Communications* 10 (4): 1232–1241.

7 Thilina, K.M., Choi, K.W., Saquib, N., and Hossain, E. (2013). Machine learning techniques for cooperative spectrum sensing in cognitive radio networks. *IEEE Journal on Selected Areas in Communications* 31 (11): 2209–2221.

8 Li, Z., Yu, F.R., and Huang, M. (2009). A distributed consensus-based cooperative spectrum-sensing scheme in cognitive radios. *IEEE Transactions on Vehicular Technology* 59 (1): 383–393.

9 Xie, S., Liu, Y., Zhang, Y., and Yu, R. (2010). A parallel cooperative spectrum sensing in cognitive radio networks. *IEEE Transactions on Vehicular Technology* 59 (8): 4079–4092.

10 Xiong, G., Kishore, S., and Yener, A. (2013). Spectrum sensing in cognitive radio networks: performance evaluation and optimization. *Physical Communication* 9: 171–183.

11 Eappen, G. and Shankar, T. (2021). Multi-objective modified grey wolf optimization algorithm for efficient spectrum sensing in the cognitive radio network. *Arabian Journal for Science and Engineering* 46 (4): 3115–3145.

12 Chen, J., Huang, S., Li, H. et al. (2019). PSO-based agent cooperative spectrum sensing in cognitive radio networks. *IEEE Access* 7: 142963–142973.

13 Tekbıyık, K., Akbunar, Ö., Ekti, A.R. et al. (2021). Spectrum sensing and signal identification with deep learning based on spectral correlation function. *IEEE Transactions on Vehicular Technology* 70 (10): 10514–10527.

14 Sheng, J., Liu, F., Zhang, Z., and Huang, C. (2022). TCVQ-SVM algorithm for narrowband spectrum sensing. *Physical Communication* 50: 101502.

15 Chen, Y., Zhang, Y., Yu, B. et al. (2021). Relay-assisted secure short-packet transmission in cognitive IoT with spectrum sensing. *China Communications* 18 (12): 37–50.

16 Eappen, G., Shankar, T., and Nilavalan, R. (2020). Efficient spectrum sensing for the relay based cognitive radio network for enhancing the network coverage for wireless patient monitoring system. *2020 Fifth International Conference on Smart and Sustainable Technologies (SpliTech)*, pp. 1–6 (23–26 September 2020). Split, Croatia: IEEE.

17 Brammya, G., Praveena, S., Ninu Preetha, N.S. et al. (2019). Deer hunting optimization algorithm: a new nature-inspired meta-heuristic paradigm. *The Computer Journal* 1–20.

18 Tian, M.W., Yan, S.R., Han, S.Z. et al. (2020). New optimal design for a hybrid solar chimney, solid oxide electrolysis and fuel cell based on improved deer hunting optimization algorithm. *Journal of Cleaner Production* 249: 119414.

19 Pang, J. and Scutari, G. (2013). Joint sensing and power allocation in nonconvex cognitive radio games: quasi-nash equilibria. *IEEE Transaction on Signal Processing* 61 (9): 2366–2382.

8

Elite Oppositional Hunger Games Search Optimization Based Cooperative Spectrum Sensing Scheme for 6G Cognitive Radio Networks

Emad A.-B. Abdel-Salam, Ayman M. Mahmoud, and Romany F. Mansour

Department of Mathematics, Faculty of Science, New Valley University, El-Kharga, Egypt

8.1 Introduction

5G cognitive radio networks (CRNs) [1] are assuming a critical part in worldwide correspondence and spectrum sharing. The primary piece of 5G CR correspondence is huge machine-type correspondence, upgraded versatile broadband, and minimum inertness ultrareliable correspondence [2]. The principle highlights of 5G are around 0.1 Gbps information rate, the pinnacle information rate is around 20 Gbps, start to finish idleness is around 1 ms, gadget availability is roughly 1 million/km^2, threefold spectrum proficiency, 500 km/h supporting versatility, traffic limit by region 10 Mbps/m^2, and a few times energy effectiveness when contrasted with 4G CRN correspondence [3]. A few fundamental advancements, for example, ultradense network, millimeter wave, and multiple input, multiple output (MIMO) are contemplated to acquire 5G CRN correspondence. Albeit, fifth age is deficient to meet the necessities of 2030. Researchers, establishments, and scientists center around 6G CRN correspondence [4].

A fundamental mark of 5G is low inertness, which requires a deterministic network to guarantee highlight point idleness with the prerequisite and accuracy of approaching future order. Creators concentrated

AI-Enabled 6G Networks and Applications, First Edition. Edited by Deepak Gupta, Mahmoud Ragab, Romany Fouad Mansour, Aditya Khamparia, and Ashish Khanna.
© 2023 John Wiley & Sons Ltd. Published 2023 by John Wiley & Sons Ltd.

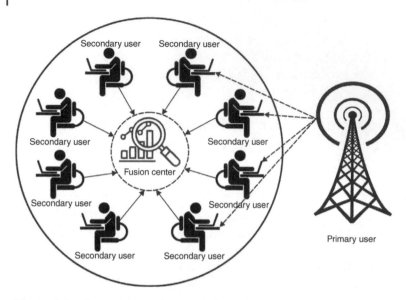

Figure 8.1 Spectrum sensing scenario.

that 6G should be competent to seek after high synchronization than 5G. Besides, there will be about greatest geological inclusion, millisecond topographical area refreshing exactness, and subcentimeter geological precision in 6G [5]. Given a primary recurrence band, the principal challenge for CR is to dependably recognize the presence of essential clients to limit the obstruction to existing interchanges. Various strategies are proposed for distinguishing the presence of sign transmission, for example, matched channel identification, energy location, including recognition methods and wavelet approach [6]. Figure 8.1 depicts the development of spectrum sensing (SS).

Nonetheless, the hidden terminal issue happens when cognitive client is shadowed, in serious multipath fading in structures with increased infiltration misfortune while essential client is working nearby. Because of the hidden terminal issue, the sensing execution for one cognitive client will be debased [7]. To forestall the hidden terminal issue, the CR network could combine the sensing aftereffects of different cognitive clients and take advantage of spatial variety among appropriated cognitive clients to upgrade the sensing dependability. Thus, a network of spatially conveyed cognitive clients, which experience different channel conditions from the objective, would have a superior possibility of identifying the essential client by trading sensing data. In this way, collaborative SS can mitigate the issue of tainted identification by taking advantage of the inherent spatial variety to lessen the likelihood of impeding essential clients [8].

Since collaborative sensing is by and large organized over a control channel, effective participation plans ought to be examined to decrease transfer speed and power prerequisites while streamlining the sensing dependability. Significant plan contemplations incorporate the upward decrease related to sensing data trade and the practicality issue of control channels. As a rule, working qualities (for example, deception versus identification probabilities) of the indicator ought to be chosen by thinking about the feasible deft throughput of cognitive clients and the likelihood of not crashing into essential clients. Collaborative SS should be possible in both brought together and decentralized ways [9]. If there should arise an occurrence of incorporated method, the choice is made by the focal combination place. The data from all the partaking clients is consolidated to settle on the choice. In collaborative sensing, the choice with regard to the client is not developed by a primary unit. Numerous metaheuristic calculations are utilized for ideal decision making for CRs [10].

This study presents elite oppositional hunger games search optimization based cooperative spectrum sensing (EOHGSO-CSS) scheme for 6G CRNs. The EOHGSO-CSS technique mainly intends to allocate the spectrum effectively in the 6G CRNs. The SS process can be carried out using different parameters such as interference, sensing time, threshold value, energy, throughput, and power allocation. Besides, the EOHSGO algorithm has been derived by the integration of elite oppositional based learning (EOBL) with traditional HGSO algorithm to improve its efficiency. For assessing the superior outcomes of the EOHGSO-CSS model, wide-ranging experiments were carried out.

8.2 Related Works

In Ref. [11], a discounted upper confidence bound (D-UCB) based cooperation partner selective technique was designed and the secondary users (SUs) learned the time varying recognition probabilities of their neighbors with the consideration of potentially maximum detection probability as cooperation partner. In Ref. [12], a new multiobjective modified GWO (MOMGWO) technique was presented for resolving the multiobjective optimized problems from the domain of SS from a CRN that is a vital paradigm from wireless communication technology. Modified in GWO was executed for balancing the trade-off among exploration as well as exploitation processes in convention GWO, for obtaining global optimum. Khaf et al. [13] proposed a scalable, partially CSS technique, which is extremely resilient to sensing data falsification (SDF) attacks. The originality of presented technique lies in partial cooperation with coalition development that decreases sensing and data sharing overhead but enhances sensing accuracy. Furthermore, this technique learned for adapting the sensing participation percentage and chooses the more rewards channel to sense for maximizing rewards but minimized energy utilization.

Le Anh et al. [14] examined the confidentiality performance of non-orthogonal multiple access (NOMA) scheme in untrusted relaying energy harvesting (UEH) network. During this network, a source communicates with user using a relaying network collected of several untrusted elements utilizing amplify-and-forward (AF) protocols. This untrusted relaying node is equipped with single antenna and utilizes the power splitting (PS) protocol for harvesting energy in received signals. In addition, for improving the confidentiality outage performance and protecting the secret data in the untrusted relaying network, the source performs as the jammer for generating artificial noise (AN).

In Ref. [15], a novel method was presented by integrating AI techniques and spectrum recognition techniques. The accuracy of outcomes is experimental by executing this novel approach to orthogonal frequency division multiplexing (OFDM) technology. Genetic algorithm (GA) was utilized for making the optimum field channel allocation and maximum accuracy to utilize spectrums. In Ref. [16], several machine learning enabled solutions are implemented for tackling the challenges of difficult sensing models from CSS to non-orthogonal several access broadcast process, comprising unsupervised learning techniques (K-means clustering and Gaussian mixture method) and supervised learning techniques (directed acyclic graph – support vector machine [SVM], k-nearest neighbour [KNN], and back propagation neural network [BPNN]). In [17], a multiagent deep reinforcement learning (MA-DRL) technique is implemented for realizing CSS from CRN. All the SUs learn an effectual sensing approach in the sensing outcomes of the number of chosen spectra for avoiding interference to primary user (PU) and for coordinating with another SU.

8.3 The Proposed Model

In this study, a new EOHGSO-CSS approach was developed to sense the spectrum effectively in the 6G CRNs. The SS process can be carried out using different parameters such as interference, sensing time, threshold value, energy, throughput, and power allocation. Besides, the EOHSGO algorithm has been derived by the integration of EOBL with traditional HGSO algorithm to improve its efficiency.

8.3.1 Design of EOHGSO Algorithm

The HGSO algorithm is an optimization method for modeling hunger and animal behavior. Hunger capability having the vital homeostatic reason for decision, behavior, and action in the animal existence symbolizes HGS. HGS arithmetical modeling starts with a population of N solutions, X, and proceeded to the objective function value for solution. It can be expressed as follows:

$$X = \begin{cases} X(t) \times (1 + rand), & r_1 < l \\ W_1 \times X_b + R \times W_2 \times |X_b - X(t)|, & r_1 > l, r_2 > E \\ W_1 \times X_b - R \times W_2 \times |X_b - X(t)|, & r_1 > l, r_2 < E \end{cases} \quad (8.1)$$

Where r_1 and r_2 are two arbitrary values, and the *rand* generates number from a standard distribution, and R denotes a parameter that value can be defined by the range $[-a, a]$ and is based on the amount of iterations in the following:

$$R = 2 \times s \times rand - s, \quad s = 2 \times \left(1 - \frac{t}{T}\right) \tag{8.2}$$

Where the variable E indicates the control variable as follows:

$$E = \mathrm{sech}(|\ Fit_i - Fit) \tag{8.3}$$

Fit_b signifies the finest value of objective function and Sech corresponding to the hyperbolic function where sech $(x) = \dfrac{2}{e^x - e^{-x}}$.

Further, W_1 and W_2 represents the hunger weight as follows:

$$W_1 = \begin{cases} H_i \times \dfrac{N}{SH} \times r_4, & r_3 < l \\ 1 & r_3 > l \end{cases} \tag{8.4}$$

$$W_2 = 2\left(1 - e^{\left(-|H_j - SH|\right)}\right) \times r_5 \tag{8.5}$$

Where r_3, r_4, and r_5 represent arbitrary number within $[0, 1]$, and the parameter SH corresponding to the solution of hunger feeling summation:

$$SH = \sum_i H_i \tag{8.6}$$

Moreover, the parameter H_i corresponding the solution hunger H_i as:

$$H_i = \begin{cases} 0, & Fit_i = Fit_b \\ H_j + H_n, & \text{otherwise} \end{cases} \tag{8.7}$$

The optimal value for the objective is provided as Fit_b, and the existing solution X_i has objective provided by Fit, and the new hunger is provided by the parameter H_n:

$$H_n = \begin{cases} LH \times (1 + r), & TH < LH \\ TH, & \text{otherwise} \end{cases} \tag{8.8}$$

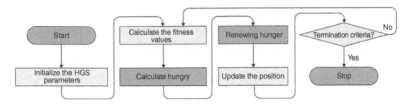

Figure 8.2 Flowchart of HGSO.

$$TH = 2\frac{Fit_i - Fit_b}{Fit_w - Fit_b} \times r_6 \times (UB - LB) \tag{8.9}$$

Fit_w provides a low value to the objective function, and $r_6 \in [0, 1]$ indicates an arbitrary parameter that indicates whether hunger has positive or harmful effects based on various aspects. Figure 8.2 demonstrates the flowchart of HGSO.

Besides, the EOHSGO algorithm has been derived by the integration of EOBL with traditional HGSO algorithm to improve its efficiency.

Evaluate the possible solution to searching space. Here, once the end condition is obtained, this approach stops, otherwise, it continues accomplishing optimal solution. EOBL is a technique utilized for enhancing the efficiency of metaheuristics [18]. Consider elite individual in existing population is $X_e = (x_{e,\,1}, x_{e,\,2}, ..., x_{e,\,D})$, for individual $X_i = (x_{i,\,1}, x_{i,\,2}, ..., x_{i,\,D})$, the elite opposition solution $\widetilde{X}_i = (\widetilde{x}_{i,1}, \widetilde{x}_{i,2}, ..., \widetilde{x}_{i,D})$ of X_i is denoted as follows:

$$\bar{x} = \eta * (da_j + db_j) - x_{e,j} \tag{8.10}$$

Where $i = 1, 2, ..., NP$, NP indicates population size, $j = 1, 2, ..., D$, $\eta \in U$ $(O, 1)$ and η shows generalized coefficient, and $[da_j, db_j]$ represent an adoptive limit of jth parameter search region and it is obtained by:

$$da_j = \min (x_{i,j}) \tag{8.11}$$

$$db_j = \max (x_{i,j}) \tag{8.12}$$

The static margin is non-conducive in saving the search experience, and thus, adoptive bound is utilized to replace the fixed bound in maintaining the search experience to make narrower opposition solution. In addition, once operator of dynamic bound makes $\tilde{x}_{i,j}$ jump out of $[da_j, db_j]$, Eq. (8.13) is employed to reset $\tilde{x}_{i,j}$:

$$\tilde{x}_{i,j} = rand\left(da_j, db_j\right) \tag{8.13}$$

The EOBL generates opposition population based elite individuals and assesses the elite and present population simultaneously. As well, it fully exploits the feature of elite individuals to encompass useful search data than standard individuals. In addition, the EOBL assists in boosting the global exploration capabilities of the HGSO method.

8.3.2 Application of EOHGSO Algorithm for CSS

To improve the efficiency of SS, several approaches are presented. Meta-heuristic algorithm has demonstrated important contribution. Here, EOHSGO algorithm is utilized for improving the SS for *CR*. The received signal comprises of N instance, and the cognitive radio receives this signal. Also, the cognitive radio is called a SU. The SUs need to make the decision regarding the existence of PU. The data is utilized by the fundamental fusion center (FC) for making the decision regarding the existence of PU. It can be considered that all the CR autonomously sense the network and communicate FC regarding their local decision [19].

The alternate and null hypotheses for the presented method are given by H_0: PU is absent.

H_1: PU is active. The optimum weight is utilized by FC to make the decision.

$$y_l(k) = \begin{cases} \varsigma_l(k) : H_0 \\ G_l s(k) + \varsigma_l(k) : H_1 \end{cases} \tag{8.14}$$

The symbols utilized in Eq. (8.14) denote $l = 1, 2, ..., M$, $\varsigma_l(k)$ signifies that the Gaussian sensing noise has a mean value of 0. $s(k)$ indicates the main signal and G_l shows the channel gain. The local signal to noise ratio (SNR) is estimated by:

$$\eta_l = \frac{\Psi_s |G_1|^2}{\sigma_l^2} \tag{8.15}$$

The signal energy is calculated by:

$$\Psi_s = \sum_{k=1}^{N} |s(k)|^2 \tag{8.16}$$

The signal energy for lth CR with N sample was estimated by:

$$\zeta_l = \sum_{k=1}^{N} |y_l(k)|^2 \tag{8.17}$$

Each received signal energy is transferred in an orthogonal manner to the FC for making decisions. The FC decided according to the outcomes of global testing statistics. The global testing statistics for lth CR is estimated by $f_l = \zeta_l + n_l$.

The test statistic is linearly estimated by $f_c = \sum_{l=1}^{M} w_l f_l = w^T f$ in which weight vector w is represented as $w \triangleq [w_1, w_2, ..., w_M]^T$ and $f = [f_1, f_2, ..., f_M]^T$. The variance f_c is given by:

$$Var(f_c H_0) = \sum_{l=1}^{M} (2N\sigma_l^4 + \vartheta_l^2) w_l^2 = w^T \chi_{H_0} w \tag{8.18}$$

Where

$$\chi_{H_0} = 2N diag^2(\sigma) + diag(\vartheta) \tag{8.19}$$

$\sigma = [\sigma_1^2, \sigma_2^2, ..., \sigma_M^2]$ and $\vartheta = [\vartheta_1^2, \vartheta_2^2, ..., \vartheta_M^2]$.

$$Var(f_c H_1) = \sum_{l=1}^{M} (2N\sigma_l^4 + \vartheta_l^2 + 4\eta_l \sigma_l^4) w_l^2 = w^T \chi_{H_1} w \tag{8.20}$$

Where

$$\chi_{H_1} = 2N diag^2(\sigma) + diag(\vartheta) + 4\Psi_s diag(G) diag(\sigma) \tag{8.21}$$

The decision rule was estimated by the energy detection at the FC as follows:

$$f_c \underset{H_1}{\overset{H_o}{\underset{<}{\overset{>}{\rule{0pt}{0pt}}}}} \tau_{fu} \tag{8.22}$$

Now τ_{fu} characterizes the applicable decision threshold.

$$\tau_{fu} = Q^{-1}(P_f)\sqrt{w^T \chi_{H_0} w} + N\sigma^2 w \tag{8.23}$$

Here P_f denotes the possibility of false alarm. Therefore, the possibility of detecting p is estimated by:

$$P_d = Q\left(\frac{Q^{-1}(P_f)\sqrt{W^T \chi_{H_0} W} - \Psi_s G^T W}{\sqrt{W^T \chi_{H_1} W}}\right) \tag{8.24}$$

The optimum weight vector (W) is obtained for maximization of ρ_d. ρ_d is maximized by reducing $\rho(W)$ while $Q(x)$ is a reduction function with respect to x.

$$\rho(W) = \frac{Q^{-1}(P_f)\sqrt{W^T \chi_{H_0} W} - \Psi_s G^T W}{\sqrt{W^T \chi_{H_1} W}} \tag{8.25}$$

$$\sum_{l=1}^{M} w_1 = 1, \quad 0 \leq w_1 \leq 1, \quad l = 1, 2, ..., M$$

8.4 Experimental Validation

This section inspects the performance validation of the EOHGSO-CSS model under several iterations and CRs.

Table 8.1 and Figure 8.3 demonstrate the maximum probability detection (MPD) outcomes of the EOHGSO-CSS model and existing techniques under several iterations. The results indicated that the EOHGSO-CSS model has outperformed the other methods with maximum MPD under all iterations. For instance, with iteration 1, the EOHGSO-CSS model has offered higher MPD of 0.8116 whereas the Manta ray foraging optimization (MRFO), dragonfly algorithm (DA), particle swarm optimization (PSO), and GA models have obtained lower MPD of 0.5605, 0.5872, 0.7697, and 0.6140, respectively. Also, with

Table 8.1 MPD analysis of EOHGSO-CSS technique under several iterations.

No. of iterations	Maximum probability detection (Pd)				
	EOHGSO-CSS	MRFO	DA	PSO	GA
1	0.8116	0.5605	0.5872	0.7697	0.6140
2	0.8627	0.6767	0.5872	0.8081	0.6268
3	0.8941	0.8813	0.7151	0.8337	0.6721
4	0.9325	0.9546	0.7070	0.8616	0.6884
5	0.9697	0.9441	0.7174	0.8860	0.7174
6	0.9778	0.9395	0.7232	0.8906	0.7151
7	0.9709	0.9592	0.7930	0.8918	0.7349
8	0.9813	0.9604	0.7930	0.8953	0.7360
9	0.9813	0.9627	0.7930	0.9255	0.7581
10	0.9825	0.9627	0.8465	0.9244	0.7732
11	0.9825	0.9627	0.9104	0.9278	0.8267
12	0.9825	0.9627	0.9104	0.9278	0.8337
13	0.9825	0.9627	0.9139	0.9313	0.8965
14	0.9825	0.9627	0.9139	0.9290	0.8965
15	0.9825	0.9627	0.9139	0.9290	0.8999
16	0.9825	0.9627	0.9139	0.9290	0.8965
17	0.9825	0.9627	0.9139	0.9290	0.8988
18	0.9825	0.9627	0.9104	0.9290	0.9069
19	0.9825	0.9627	0.9116	0.9290	0.9069
20	0.9825	0.9627	0.9092	0.9290	0.9081
21	0.9825	0.9627	0.9127	0.9290	0.9081
22	0.9825	0.9627	0.9116	0.9290	0.9081
23	0.9825	0.9627	0.9278	0.9290	0.9081
24	0.9825	0.9627	0.9255	0.9290	0.9058
25	0.9825	0.9627	0.9255	0.9290	0.9069

(Continued)

Table 8.1 (Continued)

	Maximum probability detection (Pd)				
No. of iterations	EOHGSO-CSS	MRFO	DA	PSO	GA
26	0.9825	0.9627	0.9255	0.9290	0.9151
27	0.9825	0.9627	0.9255	0.9290	0.9151
28	0.9825	0.9627	0.9255	0.9290	0.9162
29	0.9825	0.9627	0.9255	0.9290	0.9139
30	0.9825	0.9627	0.9255	0.9290	0.9151

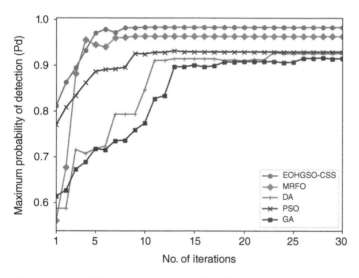

Figure 8.3 MPD analysis of EOHGSO-CSS technique under several iterations.

iteration 10, the EOHGSO-CSS model has provided increased MPD of 0.9825 whereas the MRFO, DA, PSO, and GA models have obtained lower MPD of 0.9627, 0.8465, 0.9244, and 0.7732, respectively. Moreover, with iteration 20, the EOHGSO-CSS model has offered higher MPD of 0.9825 whereas the MRFO, DA, PSO, and GA models have obtained lower MPD of 0.9627, 0.9092, 0.9290, and 0.9081, respectively. Meanwhile, with iteration 30, the EOHGSO-CSS model has offered higher MPD of 0.9825 whereas the MRFO, DA, PSO, and GA models have obtained lower MPD of 0.9627, 0.9255, 0.9290, and 0.9151, respectively.

Table 8.2 and Figure 8.4 investigate the probability of detection (PD) examination of the EOHGSO-CSS model with other methods under

Table 8.2 PD analysis of EOHGSO-CSS technique with existing approaches under *CR* = 8 and distinct PFs.

	Probability of detection (Pd); *CR* = 8				
Probability of false alarm (Pf)	EOHGSO-CSS	MRFO	DA	PSO	GA
0.05	0.2796	0.1848	0.0387	0.0028	0.0464
0.10	0.5795	0.4846	0.4334	0.5128	0.2360
0.15	0.7999	0.7307	0.6871	0.5359	0.4641
0.20	0.8409	0.7819	0.7358	0.7409	0.5974
0.25	0.8639	0.8127	0.7947	0.7973	0.6512
0.30	0.9178	0.8563	0.8229	0.8486	0.6461
0.35	0.9485	0.9024	0.8665	0.8588	0.6794
0.40	0.9690	0.9383	0.9203	0.9178	0.7409
0.45	0.9690	0.9152	0.9101	0.9383	0.8434
0.50	0.9818	0.9203	0.9280	0.9229	0.8255
0.55	0.9921	0.9280	0.9537	0.9690	0.7999
0.60	0.9982	0.9511	0.9562	0.9767	0.8563
0.65	0.9982	0.9511	0.9767	0.9665	0.8845
0.70	0.9982	0.9742	0.9870	0.9818	0.8691
0.75	0.9982	0.9947	0.9742	0.9870	0.9126

(*Continued*)

Table 8.2 (Continued)

Probability of false alarm (Pf)	Probability of detection (Pd); *CR* = 8				
	EOHGSO-CSS	MRFO	DA	PSO	GA
0.80	0.9982	0.9844	0.9870	0.9998	0.9460
0.85	0.9982	0.9921	0.9972	0.9947	0.9921
0.90	0.9982	0.9921	0.9972	0.9947	0.9742
0.95	0.9982	0.9972	0.9921	0.9947	0.9818
1.00	0.9982	0.9972	0.9895	0.9947	0.9870

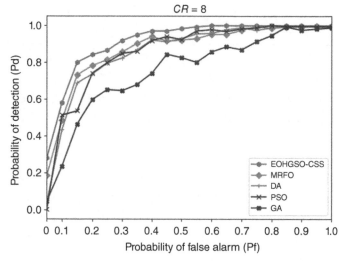

Figure 8.4 PD analysis of EOHGSO-CSS technique under *CR* = 8 and distinct PFs.

CR = 8 and distinct probability of false alarms (PFs). The results indicated that the EOHGSO-CSS model has accomplished enhanced PD under all PFs. For instance, with PF of 0.05, the EOHGSO-CSS model has reached higher PD of 0.2796 whereas the MRFO, DA, PSO, and

GA models have resulted in reduced PD of 0.1848, 0.0387, 0.0028, and 0.0464, respectively. Along with that, with PF of 0.25, the EOHGSO-CSS technique has reached higher PD of 0.8639 whereas the MRFO, DA, PSO, and GA models have resulted in reduced PD of 0.8127, 0.7947, 0.7973, and 0.6512, respectively. Moreover, with PF of 0.65, the EOHGSO-CSS model has reached higher PD of 0.9982 whereas the MRFO, DA, PSO, and GA systems have resulted in reduced PD of 0.9511, 0.9767, 0.9665, and 0.8845 correspondingly. Furthermore, with PF of 1.00, the EOHGSO-CSS algorithm has reached higher PD of 0.9982 whereas the MRFO, DA, PSO, and GA models have resulted in reduced PD of 0.9972, 0.9895, 0.9947, and 0.9870 correspondingly.

Table 8.3 and Figure 8.5 inspect the PD examination of the EOHGSO-CSS technique with other methods under $CR = 10$ and distinct PFs. The

Table 8.3 PD analysis of EOHGSO-CSS technique with existing approaches under $CR = 10$ and distinct PFs.

	Probability of detection (Pd); $CR = 10$				
Probability of false alarm (Pf)	**EOHGSO-CSS**	**MRFO**	**DA**	**PSO**	**GA**
0.05	0.7462	0.7145	0.7069	0.5051	0.5242
0.10	0.8376	0.8134	0.8046	0.7576	0.7259
0.15	0.8693	0.8388	0.8236	0.8173	0.7297
0.20	0.9124	0.8870	0.8756	0.8832	0.8033
0.25	0.9492	0.9187	0.9124	0.9048	0.8173
0.30	0.9746	0.9467	0.9073	0.9086	0.8350
0.35	0.9911	0.9530	0.9213	0.9403	0.8490
0.40	0.9911	0.9543	0.9581	0.9492	0.8896
0.45	0.9923	0.9847	0.9632	0.9657	0.9264
0.50	0.9961	0.9771	0.9657	0.9670	0.9048
0.55	0.9961	0.9835	0.9670	0.9670	0.9314

(*Continued*)

Table 8.3 (Continued)

Probability of false alarm (Pf)	Probability of detection (Pd); $CR = 10$				
	EOHGSO-CSS	MRFO	DA	PSO	GA
0.60	0.9974	0.9835	0.9885	0.9847	0.9327
0.65	0.9987	0.9898	0.9809	0.9822	0.9441
0.70	0.9999	0.9961	0.9885	0.9898	0.9441
0.75	0.9987	0.9961	0.9911	0.9923	0.9517
0.80	0.9987	0.9961	0.9911	0.9961	0.9733
0.85	0.9999	0.9961	0.9949	0.9949	0.9873
0.90	0.9999	0.9949	0.9974	0.9949	0.9822
0.95	0.9999	0.9984	0.9974	0.9987	0.9898
1.00	0.9999	0.9991	0.9989	0.9988	0.9936

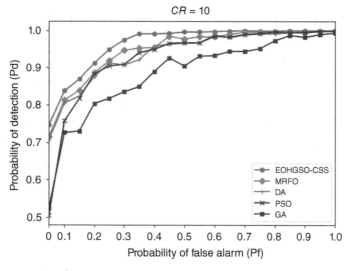

Figure 8.5 PD analysis of EOHGSO-CSS technique under $CR = 10$ and distinct PFs.

results indicated that the EOHGSO-CSS method has accomplished enhanced PD under all PFs. For instance, with PF of 0.05, the EOHGSO-CSS approach has reached higher PD of 0.7462 whereas the MRFO, DA, PSO, and GA models have resulted in reduced PD of 0.7145, 0.7069, 0.5051, and 0.5242 correspondingly. Likewise, with PF of 0.25, the EOHGSO-CSS model has reached higher PD of 0.9492 whereas the MRFO, DA, PSO, and GA systems have resulted in reduced PD of 0.9187, 0.9124, 0.9048, and 0.8173, respectively. Followed by, with PF of 0.65, the EOHGSO-CSS model has reached higher PD of 0.9987 whereas the MRFO, DA, PSO, and GA methods have resulted in reduced PD of 0.9898, 0.9809, 0.9822, and 0.89441 correspondingly. Finally, with PF of 1.00, the EOHGSO-CSS technique has reached higher PD of 0.9999 whereas the MRFO, DA, PSO, and GA systems have resulted in reduced PD of 0.9991, 0.9989, 0.9988, and 0.9936, respectively.

Table 8.4 and Figure 8.6 demonstrate the PD examination of the EOHGSO-CSS system with other methods under $CR = 12$ and distinct PFs. The results demonstrated that the EOHGSO-CSS method has accomplished enhanced PD under all PFs. For instance, with PF of 0.05, the EOHGSO-CSS methodology has reached superior PD of 0.8277 whereas the MRFO, DA, PSO, and GA techniques have resulted in reduced PD of 0.7892, 0.6505, 0.5027, and 0.5130 correspondingly. Similarly, with PF of 0.25, the EOHGSO-CSS approach has reached increased PD of 0.9832 whereas the MRFO, DA, PSO, and GA techniques have resulted in reduced PD of 0.9318, 0.9074, 0.9010, and 0.8239 correspondingly. In addition, with PF of 0.65, the EOHGSO-CSS approach has reached higher PD of 0.9986 whereas the MRFO, DA, PSO, and GA techniques have resulted in reduced PD of 0.9922, 0.9896, 0.9832, and 0.9472 correspondingly. Lastly, with PF of 1.00, the EOHGSO-CSS algorithm has reached maximal PD of 0.9999 whereas the MRFO, DA, PSO, and GA models have resulted in reduced PD of 0.9986, 0.9982, 0.9973, and 0.9845 correspondingly.

From the aforementioned results, it can be apparent that the EOHGSO-CSS technique has accomplished maximum outcomes over the other methods.

Table 8.4 PD analysis of EOHGSO-CSS technique with existing approaches under CR = 12 and distinct PFs.

| Probability of false alarm (Pf) | Probability of detection (Pd); CR = 12 | | | | |
	EOHGSO-CSS	MRFO	DA	PSO	GA
0.05	0.8277	0.7892	0.6505	0.5027	0.5130
0.10	0.8920	0.8213	0.8162	0.7841	0.6672
0.15	0.9459	0.8804	0.8573	0.8521	0.7853
0.20	0.9729	0.9408	0.8406	0.8290	0.7918
0.25	0.9832	0.9318	0.9074	0.9010	0.8239
0.30	0.9870	0.9318	0.9241	0.9267	0.8380
0.35	0.9935	0.9716	0.9485	0.9472	0.8560
0.40	0.9922	0.9652	0.9536	0.9498	0.8894
0.45	0.9922	0.9652	0.9601	0.9716	0.9138
0.50	0.9960	0.9870	0.9909	0.9845	0.9318
0.55	0.9935	0.9857	0.9755	0.9780	0.9369
0.60	0.9999	0.9935	0.9883	0.9819	0.9344
0.65	0.9986	0.9922	0.9896	0.9832	0.9472
0.70	0.9999	0.9947	0.9896	0.9896	0.9523
0.75	0.9999	0.9986	0.9947	0.9909	0.9665
0.80	0.9986	0.9986	0.9922	0.9909	0.9870
0.85	0.9986	0.9960	0.9935	0.9883	0.9909
0.90	0.9986	0.9960	0.9947	0.9896	0.9883
0.95	0.9999	0.9960	0.9960	0.9870	0.9896
1.00	0.9999	0.9986	0.9982	0.9973	0.9845

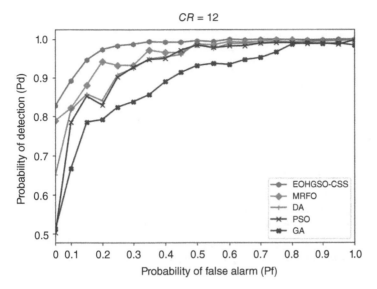

Figure 8.6 PD analysis of EOHGSO-CSS technique under $CR = 12$ and distinct PFs.

8.5 Conclusion

In this study, a novel EOHGSO-CSS approach has been developed to sense the spectrum effectively in the 6G CRNs. The SS process can be carried out using different parameters such as interference, sensing time, threshold value, energy, throughput, and power allocation. Besides, the EOHSGO algorithm has been derived by the integration of EOBL with traditional HGSO algorithm to improve its efficiency. For assessing the superior outcomes of the EOHGSO-CSS model, wide-ranging experiments were carried out. The extensive comparative study highlighted the enhancements of the EOHGSO-CSS model over the other approaches in terms of different measures. Therefore, the EOHGSO-CSS technique can be employed as a proficient tool to sense spectrum in the 6G enabled CRNs. In future, deep learning models can be employed for optimal CSS in the 6G enabled CRNs.

References

1 Gul, N., Ahmed, S., Kim, S.M., and Kim, J. (2021). Robust spectrum sensing against malicious users using particle swarm optimization. *ICT Express* https://doi.org/10.1016/j.icte.2021.12.008.

2 Singh, K.K., Yadav, P., Singh, A. et al. (2021). Cooperative spectrum sensing optimization for cognitive radio in 6 G networks. *Computers and Electrical Engineering* 95: 107378.

3 Eappen, G. and Shankar, T. (2020). A survey on soft computing techniques for spectrum sensing in a cognitive radio network. *SN Computer Science* 1 (6): 1–36.

4 Eappen, G., Shankar, T., and Nilavalan, R. (2020). Efficient spectrum sensing for the relay based cognitive radio network for enhancing the network coverage for wireless patient monitoring system. *2020 Fifth International Conference on Smart and Sustainable Technologies (SpliTech)*, pp. 1–6 (23–26 September 2020). Split, Croatia: IEEE.

5 Sheng, J., Liu, F., Zhang, Z., and Huang, C. (2022). TCVQ-SVM algorithm for narrowband spectrum sensing. *Physical Communication* 50: 101502.

6 Wan, S., Hu, J., Chen, C. et al. (2020). Fair-hierarchical scheduling for diversified services in space, air and ground for 6G-dense internet of things. *IEEE Transactions on Network Science and Engineering* 8 (4): 2837–2848.

7 Pham, Q.V., Nguyen, D.C., Mirjalili, S. et al. (2021). Swarm intelligence for next-generation networks: recent advances and applications. *Journal of Network and Computer Applications* 191: 103141.

8 Qamar, F., Siddiqui, M.U.A., Hindia, M.H.D. et al. (2020). Issues, challenges, and research trends in spectrum management: a comprehensive overview and new vision for designing 6G networks. *Electronics* 9 (9): 1416.

9 Ivanov, A., Tonchev, K., Poulkov, V., and Manolova, A. (2021). Probabilistic spectrum sensing based on feature detection for 6G cognitive radio: a survey. *IEEE Access* 9: 116994–117026.

10 Song, Z., Gao, Y., and Tafazolli, R. (2021). A survey on spectrum sensing and learning technologies for 6G. *IEICE Transactions on Communications* 104 (10): 1207–1216.

11 Ning, W., Huang, X., Yang, K. et al. (2020). Reinforcement learning enabled cooperative spectrum sensing in cognitive radio networks. *Journal of Communications and Networks* 22 (1): 12–22.

12 Eappen, G. and Shankar, T. (2021). Multi-objective modified grey wolf optimization algorithm for efficient spectrum sensing in the cognitive radio network. *Arabian Journal for Science and Engineering* 46 (4): 3115–3145.

13 Khaf, S., Alkhodary, M.T., and Kaddoum, G. (2021). Partially cooperative scalable spectrum sensing in cognitive radio networks under SDF attacks. *IEEE Internet of Things Journal* 9: 8901–8912.

14 Le Anh, T. and Hong, I.P. (2020). Secrecy performance of a multi-NOMA-MIMO system in the UEH relaying network using the PSO algorithm. *IEEE Access* 9: 2317–2331.

15 Yilmazel, R. and Inanç, N. (2021). A novel approach for channel allocation in OFDM based cognitive radio technology. *Wireless Personal Communications* 120 (1): 307–321.

16 Shi, Z., Gao, W., Zhang, S. et al. (2020). Machine learning-enabled cooperative spectrum sensing for non-orthogonal multiple access. *IEEE Transactions on Wireless Communications* 19 (9): 5692–5702.

17 Zhang, Y., Cai, P., Pan, C., and Zhang, S. (2019). Multi-agent deep reinforcement learning-based cooperative spectrum sensing with upper confidence bound exploration. *IEEE Access* 7: 118898–118906.

18 Nguyen, H. and Bui, X.N. (2021). A novel hunger games search optimization-based artificial neural network for predicting ground vibration intensity induced by mine blasting. *Natural Resources Research* 30 (5): 3865–3880.

19 Khanduja, N. and Bhushan, B. (2021). Chaotic state of matter search with elite opposition based learning: a new hybrid metaheuristic algorithm. *Optimal Control Applications and Methods* https://doi.org/10.1002/oca.2810.